Advanced Biodiesel - Technological Advances, Challenges, and Sustainability Considerations

Edited by Islam Md Rizwanul Fattah

Published in London, United Kingdom

Advanced Biodiesel - Technological Advances, Challenges, and Sustainability Considerations
http://dx.doi.org/10.5772/intechopen.104057
Edited by Islam Md Rizwanul Fattah

Contributors
P. Bharathi, R. Dayana, G.M. Shanthini, Manju D. Tanwar, Pankaj K. Tanwar, Yashas Bhand, Sarang Bhand, Kiran Jadhav, Suhas Bhand, Welela Meka Kedir, Issis Claudette Romero-Ibarra, Araceli Martínez Ponce Escuela, Wendy Eridani Medina Muñoz, Gabriela Elizabeth Mijangos Zúñiga, Rajeshwar Rao Gandhe, Ajin Sekhar, Rajkumar Muthu, Aqsa Parvaiz, Muhammad Saeed, Ghulam Mustafa, Faiz Ahmad, Aneela Shadab, Timothy J. Tse, Farley Chicilo, Martin J.T. Reaney, Venkatesh Meda, Li Zhou, Anshebo Getachew Alemu, Teketel Alemu

First published in London, United Kingdom, 2024 by IntechOpen
IntechOpen is the global imprint of INTECHOPEN LIMITED, registered in England and Wales, registration number: 11086078, 5 Princes Gate Court, London, SW7 2QJ, United Kingdom

British Library Cataloguing-in-Publication Data
A catalogue record for this book is available from the British Library

Additional hard and PDF copies can be obtained from orders@intechopen.com

Advanced Biodiesel - Technological Advances, Challenges, and Sustainability Considerations
Edited by Islam Md Rizwanul Fattah
p. cm.
Print ISBN 978-1-83768-137-2
Online ISBN 978-1-83768-138-9
eBook (PDF) ISBN 978-1-83768-139-6

For EU product safety concerns:
IN TECH d.o.o., Prolaz Marije Krucifikse Kozulić 3, 51000 Rijeka, Croatia,
info@intechopen.com or visit our website at intechopen.com.

Meet the editor

Dr. Islam Md Rizwanul Fattah is a research fellow in the School of Civil and Environmental Engineering, Faculty of Engineering and IT, University of Technology Sydney (UTS), Australia, where he explores the effective use of waste for sustainable energy applications. His research interests stem from his Ph.D. in the combustion of renewable fuels, which he completed in 2019 at the University of New South Wales (UNSW), Sydney, Australia, and are focused on reducing emissions (primarily PM/soot) from diesel combustion. In 2014, he earned an MEng from the University of Malaya (UM) and a BSc in Mechanical Engineering from the Bangladesh University of Engineering and Technology (BUET). Dr. Fattah was named a highly cited researcher by Clarivate Analytics in 2022 and listed among the top 2% of scientists in the world by Standford University for three consecutive years (2020–2022). He has published more than 120 articles and is an associate editor for *Frontiers in Energy Research*, *Frontiers in Catalysis*, and *Journal of Energy and Power Technology*. He is managing and has managed more than fifteen special journals and edited more than fifty articles. Throughout his career, he also has peer-reviewed more than 350 journal articles. Dr. Fattah would like to remain at the forefront of research in the field of renewable and sustainable energy.

Contents

Preface

In an era marked by escalating concerns about environmental sustainability and the relentless pursuit of alternative energy sources, the subject of biodiesel has gained significant prominence. It is in this context that we proudly present *Advanced Biodiesel – Technological Advances, Challenges, and Sustainability Considerations*. This comprehensive volume delves into the intricate world of biodiesel production, refining, and application, offering readers an in-depth exploration of this vital branch of renewable energy.

Biodiesel, derived from natural sources, not only serves as a cleaner and more environmentally friendly fuel alternative but also stands as a beacon of hope for reducing our dependence on fossil fuels. This book serves as a testament to the dynamic progress in biodiesel technology and the relentless efforts of researchers, scientists, and industry experts who are continually striving to refine and expand the horizons of this promising field.

Within these pages, the reader embarks on a fascinating exploration of the technological advancements that have elevated biodiesel production and usage to new heights. The chapters in this book are a testament to the depth and breadth of biodiesel research. They span a diverse range of topics, from innovative catalytic processes for biodiesel production to the exploration of cutting-edge nanomaterials and biotechnological interventions. Our expert contributors share their insights into the challenges and opportunities in biodiesel refining, applications, and sustainability considerations, offering a multifaceted perspective on this multifaceted field.

As you embark on this journey through the chapters of *Advanced Biodiesel – Technological Advances, Challenges, and Sustainability Considerations*, we hope that you will gain a profound understanding of the strides made in the biodiesel industry and the myriad challenges and opportunities that lie ahead. This book stands as a testament to the shared dedication of experts from around the world who are working tirelessly to create a more sustainable and environmentally responsible energy future.

Happy reading, and onward to a brighter and more sustainable future!

Islam Md Rizwanul Fattah
School of Civil and Environmental Engineering,
Faculty of Engineering,
University of Technology Sydney,
Ultimo, Australia

Section 1

Biodiesel Introduction

Chapter 1

Biofuels: Production and Properties as Substitute Fuels

Manju D. Tanwar, Pankaj K. Tanwar, Yashas Bhand, Sarang Bhand, Kiran Jadhav and Suhas Bhand

Abstract

Renewable sources include plants and animal fats, which are the main components of biofuels. Biofuels are free from sulfur, aromatics, metals, and crude oil residues. Since biofuels are more lubricating than petroleum diesel fuel, they are nonflammable and extend the life of diesel engines. As a result of this study, the main chemical and physical properties of biofuels were investigated, including their lubricity, viscosity, calorific value, and cetane number, which indicate the quality of renewable fuels, and compared with the other. We examined and compared the combustion characteristics of various types of biofuels as an alternative fuel, as well as their emissions characteristics. Biodiesel and biodiesel blends are compared to mineral diesel, as well as their performance in CI engines in this study's review. With modified combustion equipment, biodiesel fuels can potentially reduce air pollution in diesel engines and are a very good substitute for fossil fuels. There is a need for more research and technological development in order for biofuels to become economically viable. Biofuel/biodiesel research should therefore be supported with policies that make their prices competitive with other conventional sources of energy. In the current state of affairs, biofuels are more effective when used alongside other sources of energy.

Keywords: biodiesel, alternative fuel, physical and chemical properties, emission, compression ignition engine

1. Introduction

World Energy Resources 2013 reported that 82% of electricity in 2013 was generated from fossil fuels, 13% from renewables, and the rest from nuclear sources [1]. Hydroelectric, wind, and solar power generate large amounts of power, but oil reserves are diminishing and could disappear within a century [2], making it essential to find replacements for petroleum-based fuels. A large portion of petroleum-based products are consumed by the transportation sector, which demands approximately 38% [3]. Scientists are also searching for alternative fuel sources due to the high price of fossil fuels, harmful greenhouse gas emissions, and the diversity of energy production. However, diesel engines still contribute significantly to greenhouse gas emissions and adverse effects on human health, despite better fuel economy and lower car taxes. There has been some research into solar cars as a solution to this issue, but their high cost and inconsistency make them unsuitable for daily use. Fuel and lubricant alternatives have been sought as a result [4]. Alternative sources of energy, such as solar, can still be utilized in a variety of ways. During photosynthesis, solar rays are converted into stored energy within plant tissue, which helps the plant live and reproduce through seed production. A biofuel can be produced using this energy change phenomenon. For compression ignition engines, biofuel produced from renewable sources is considered the best alternative to mineral diesel fuel.

The transportation industry contributes significantly to a country's socioeconomic growth and development. Individuals' quality of life is measured by the ease of moving goods and services. Transportation services are affordably and safely provided by governments across jurisdictions. Over 90% of the fossil fuel products are consumed in the transportation sector [5, 6]. By 2030, on-road transport will consume 50% of total energy, and by 2050, 80% [7]. Transport sector energy consumption in 2015 included passenger vehicles (cars and bikes), buses, air, passenger rail, and airline freight and heavy trucks, light trucks, and marine transport consume 35% of the energy used in the transportation sector. By 2050, liquid biofuel consumption will increase to 652 billion liters, and biodiesel consumption in the transport sector will rise to 180 billion liters [8]. By 2035, there will be more than 2 billion cars on the road, and this number will rise to 2.5 billion by 2050 [9, 10]. Environmental consequences and costs will be unimaginable if these cars run on fossil fuels. Low-carbon transport systems include biofuels, hydrogen, and electric vehicles (EVs). The use of ICEs will remain important in most developing countries for some time to come, despite the fact that hydrogen and electric vehicles avoid land use and impact air quality [11].

Most of investigations have been done based on use of the fuels in the diesel engine without any modification. Based on the so many literatures and studies of previous researchers, the authors have attempted to review important research on biodiesel production processes, biodiesel physical and chemical properties, and its performance and emission characteristics as compression ignition engine's fuels.

Because of global warming and increasing air pollution, alternative fuels are increasingly being used in IC engines. Among the alternatives, biodiesel fuels seem remarkably interesting. They will be produced during a renewable way and possess certain advantageous properties that give them the potential to lower pollutants and CO_2 emissions from IC engines.

The review deals with the status of biodiesel fuels and tries to elaborate the future direction for more wider utilization and the possible roles of biodiesel fuels in attaining the far-reaching goal of low-carbon economy using sustainable energy resources.

2. Biodiesel production

There are several options for biomass-derived fuels production involving chemical, biological, and thermochemical processes. An extensive map for these pathways can be seen in **Figure 1** [12]. Two of the most promising fuels appear to be biodiesel or synthetic fuels such, as Fischer-Tropsch diesel [13]. This is because other potential fuels, such as Ethanol, Methanol, and LPG, do not perform as well in modern engines. Within IC engines, there are biofuel options for Gasoline (SI) and Diesel (CI) engines.

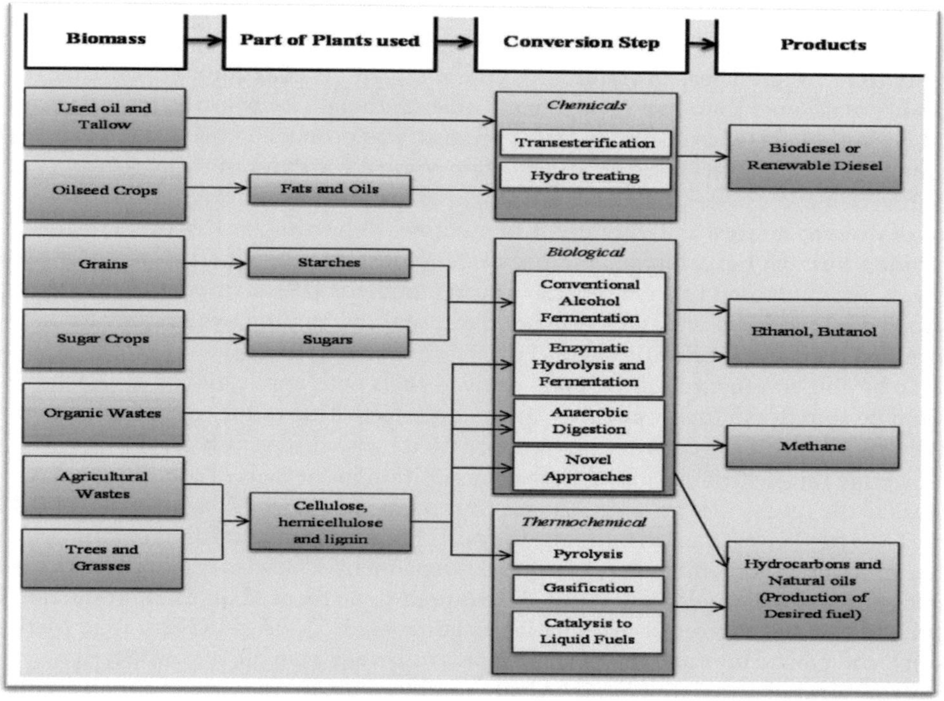

Figure 1.
An image showing biofuel pathways from feedstock to products [12].

Biogas and primary alcohols are the main fuels for SI engine. Ethanol from sugarcane was a key player in Brazil in the 1970s following the oil crisis, unfortunately the recent low petrol prices have undermined the green transport program decreasing the advantages. The government has overhauled their economic policy to ensure that the mandatory mix of ethanol in petrol is increased by 2.5% and have reinstated a levy on fossil fuels. A lot of current research is aimed toward diesel alternatives, as CI engines boast better fuel economy and lower car tax than petrol engines. An extensive map of pathways for the creation of all biofuels can be seen below in **Figure 1** taken from IRENE Transport sector summary charts, 2014 [12].

As shown in **Figure 1**, vegetable oils are a source of biomass, which has the advantages over other energy sources of not exhausting the soil or damaging the environment [6]. Crude vegetable oils can be run through modern engines though their viscosity, calorific value, and freezing point are inferior to diesel fuel.

In literature, there are many examples of fuel properties for biodiesel created by transesterification, which involves removing the glycerides and combining oil esters of vegetable oil with alcohol. This creates fuels made up of alkyl esters of long chain fatty acids, usually found to be fatty acid methyl esters (FAMEs). The major cost in its production is the enzymatic reaction, which, due to the environment-friendly and less energy-intensive nature, prefers enzymes over chemical catalysts [14]. Hwang et al. found that producing biodiesel from Waste Cooking Oil (WCO) through transesterification with methanol and a Sodium Methoxide (NaOCH3) catalyst actually reduced the production of Carbon Monoxide (CO), Hydrocarbons (HC), and Particulate Matter (PM) at low loads, compared to diesel at conventional operating conditions [15]. A more recent development in transesterification and hydrogenation is to create biofuels from algae lipids. Alga is interesting due to its high production rates, lipid content, and rapid growth cycles [16]. They also do not compete for land growth and have the ability to grow anywhere in water [17]. However, Viêgas et al. found that oxidative stability was lower than that of soybean biodiesel; though by working on Palladium and Nickel catalysts, an optimized final product was produced [18]. Another technique of producing fuel to be examined from **Figure 1** is Pyrolysis. The thermochemical reaction takes place at high temperatures in the absence of oxygen. One way to do this of present interest in Literature is by microwave heating, as this process of warming through radiation decreases the energy lost, increasing efficiency and economy, when compared to conventional heating methods [19]. Krutof et al., looked at pyrolysis oils and fish oils and the possibilities within blending, which they found increased their fuels' calorific value [20].

To be able to synchronize biomass-derived fuels with applications for which they would be suited, the fuels must first be characterized. This is due to the high potential of differentiation between the chemical products, with different methods of creation and a huge range variable affecting the composition immensely. To start you may organize the feedstocks for biofuels into four main categories [7] as seen in **Tables 1** and **2**. To further categorize biofuels, they can be organized according technology of their production, on which they are greatly dependent. These are enlisted in **Table 3**. First-generation biofuels are potentially required to be focused on food production commodities and as a result may not be sustainable [3, 7]. FAME (Fatty Acid Methyl Ester), the most common biofuel in Europe, from vegetable oil, is considered a first-generation biofuel as it is exclusively produced using transesterification technology [21]. However, second-generation advanced biofuels, produced from Fischer-Tropsch synthesis, in addition to hydrothermal, pyrolysis liquefaction, and alternative catalytic procedures, maybe sustainable for future societies with their advantages as eco-friendly fuels [3, 7, 23]. These aim to overcome the limitations of first-generation biofuels [21], but they are still deeply reliant on the price of feedstock. Previous

Category	Edible vegetable oil	Non-edible vegetable oil	Waste or recycled oil	Animal fats
Examples	Canola, soybean, peanut, sunflower, palm and coconut oil	Jatropha curcas, Calophyllum inophyllum, Moringa oleifera and Croton megalocarpus	Waste or recycled oils both edible and nonedible	Chicken fat, pork lard, beef tallow and poultry fat

Table 1.
A table showing categories of biofuel feedstocks [7].

First generation	Second generation	Third generation	Fourth generation
Feedstock: sugar, starch, vegetable oils or animal fats	**Feedstock:** non-food crops, wheat straw, corn, wood, solid waste, and energy crop.	**Feedstock:** algae	**Feedstock:** genetically modified algae (Cyanobacteria)
Examples: bio-alcohols, vegetable oil, biodiesel, bio-syngas and biogas	**Examples:** bio alcohols, bio-oil, DMF, biohydrogen and bio-Fischer-Tropsch diesel	**Example:** biodiesel	**Example:** Biodiesel

Table 2.
A table showing biofuel classification on different generation technologies [3].

Advantage	Explanation
Biodiesel contains more energy than diesel	Biodiesel carries 4.5 units of energy against each unit of fossil fuels [3]. Also, plants can be harvested completely by modern technology and the tops could be burned in the high efficiency distilleries or boilers, which can be used for gasification and could increase the total energy return up to 40% [7]
Biodiesel decreases the reliance on petroleum [3]	Biodiesel provides and alternative which means when oil runs out we will not be solely dependent on it which will positively impact on the transportation market [3]
Biodiesel can be produce locally from renewable sources	Biodiesel is possible to manufactured using edible, non-edible, waste oils, fats and oil seeds [3, 7, 11, 21]
Biodiesel is biodegradable, non-toxic and non-flammable [3, 9, 13, 22]	Bio-diesel degrades four times faster than diesel [9]. The higher flash point makes the storage safer [9]
Biodiesel can be used in Compression Ignitions engines without need for modification [9, 11, 13, 14].	Biodiesel will give and efficient performance in existing diesel engines [14] with no substantial modifications to the engine [9]
Biodiesel is environmentally friendly by emitting less pollutants [11, 22]	Biodiesel has the potential to be a carbon neutral fuel [9, 21], due to balance between the amount of CO2 emissions and the amount of CO2 absorbed by the plants producing vegetable oil being equal [9]. Also in general it has reduced greenhouse gas (GHGs) emissions [14] by 78% when compared to petroleum diesel [13]. Biodiesel contains no sulfur [9, 13] and negligible aromatic content [13]. However, with a high concentration of oxygen, the fuel produces significant reductions of sulfur dioxide (SO2), soot, carbon monoxide (CO) and unburned hydrocarbons (UHC) [13].
Biodiesel is safer	The above characteristics of biodiesel make it non-toxic and therefore safer to breathe [4]. Also biodiesel gives a 90% reduction in cancer risks, according to Ames mutagenicity tests [9].

Table 3.
A table showing the advantages of biodiesel in literature.

studies show that the feedstock cost embodies 75–80% of the total production cost [8]. In addition, the sources of the feedstock vary country to country and according to environmental conditions. For example, soybean for North America, sunflower and rapeseed for Europe, palm for Southeast Asia, coconut for tropic and sub-tropic areas, etc. [4]. Therefore, choosing the source with the highest oil yield is detrimental to success of producing low-cost biodiesel [8].

2.1 Fischer-Tropsch synthesis

Chemical, biological, and thermochemical processes can be used to produce biomass-derived fuels. Biodiesel and synthetic fuels such as Fischer-Tropsch diesel appear to be the most promising fuels [13, 24]. Modern engines do not perform well with other potential fuels, such as ethanol, methanol, and LPG. Furthermore, biomass will play an important role soon as one of the most important renewable energy sources. Another significant research area in the field of renewable energy includes the use of hydrogen [25, 26]. Using Fischer-Tropsch as a substitute for liquid fossil fuels is considered a good solution for the abovementioned issue [27–29].

$$n(CO + 2H_2) \rightarrow (CH_2)_n + n(H_2O) \tag{1}$$

During FTS, CO and H_2 react on the catalyst surface, resulting in a surface polymerization reaction. The process is described as the formation of a carbide layer on the surface of catalysts, discovered nine decades ago by Fischer and Tropsch [30–33]. Hydrocarbon molecules that are released from catalyst surfaces and reabsorbed undergo two further reactions. The reaction shown in Eqs. (2) and (3) is the reaction of methane formation (as an unwanted product in FTS); it is considered irreversible. FTS is also characterized by the water-gas shift reaction, which produces water as a coproduct; this reaction plays a significant role in reactors where the reaction occurs over cobalt catalysts and produces carbon dioxide [26, 31, 33, 34].

$$(CO + 3H_2) \rightarrow (CH_4) + (H_2O) \tag{2}$$

$$(CO + H_2O) \leftrightarrow (CO_2) + (H_2) \tag{3}$$

Fischer-Tropsch is an alternative catalytic procedure involved in the Biomass to Liquid (BTL) fuel route, which results from the FT-synthesis technology pioneered by the Germans in the 1920s [21]. The original scheme utilized coal in the place of biomass, as required during WWII. Another FT route is the Gas to Liquid path by way of converting Natural gas. The modern process is a set of chemical reactions of synthesizing hydrocarbons from a mixture of carbon monoxide and hydrogen [15], otherwise referred to as syngas, which was produced through a gasification process at the previous stage. An overview of the Fischer-Tropsch process is outline in the flow chart in **Figure 2** [35].

2.2 Vegetable oils and animal fats

Other biodiesel production techniques include ultrasonic cavitation, hydrodynamic cavitation, microwave irradiation, response surface technology, two-step reaction process, etc. [36]. Vegetable oil or animal fats are another alternative energy source, which appear to be an excellent substitute for transportation fuel. This is due to the ease of their production, utilization, storage, and the significant reduction achievable in pollutant emissions, mainly carbon dioxide (CO_2) [13]. Some examples of types of oilseed crop whose produce can be applicable are Canola, Palm, Corn, Cotton, Crambe, Linseed, Peanut, Rapeseed, Safflower, Sesame, Soya bean, Sunflower, Palm, Babassu, and Karanja, which are all examined in this paper. The main problems of using crude

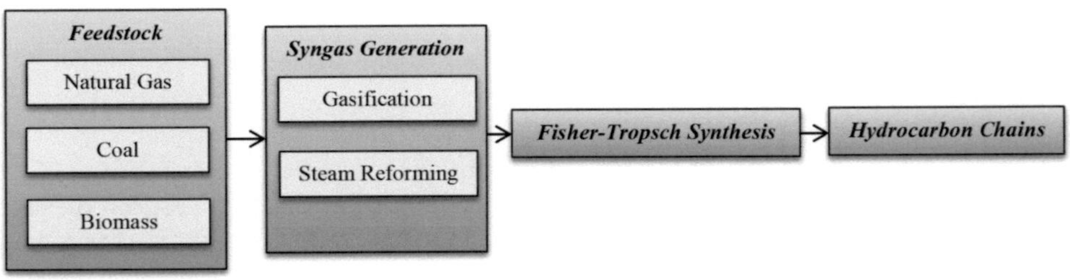

Figure 2.
An overview of Fischer-Tropsch process.

vegetable oils, which make it inferior as a fuel, are their viscosity, heating value, freezing point, etc. [4]. Chemical treatment such as alkali-catalyzed transesterification can be used to improve properties and produce biodiesel [4, 22]. Biodiesel is made up of alkyl esters of long-chain fatty acids with the major cost in its production being in the enzymatic reaction, which, due to the environment friendly and less energy-intensive nature, prefers enzymes over chemical catalysts [14].

2.3 Advantages of biofuels

Biomass is constantly being converted into energy by ways of natural processes such as aerobic digestion, fermentation, and composting. For example, by merely consuming vegetables and plant life, we in turn are converting biomass into the energy we require to live. By taking this primal concept and applying it literally to the creation of biofuels, it could be possible to devise a new mainstream renewable energy solution. Vegetable oils have the advantages of being available across the globe. They are renewable, as they produce oil seeds that can be planted to further the biofuel cycle and are "greener" as they seldom contain sulfur. Using these oils to then create biodiesel implements dependencies such as the viability of the fuel. This will rely on such factors as availability of the raw material in commercial quantity, ease of oil extraction, the oil content (yield) of the plant seeds as well as the quality of their product, which will need to meet the basic fuel characteristics for diesel fuels [22]. Extensive advantages of biodiesel are discussed in **Table 1** and were discovered through research.

2.4 Disadvantages of biofuels

There are several drawbacks to using crude vegetable oil, and these can be seen below and were taken from Gandure et al. [22]. These problems contribute to the reasoning behind chemically modifying the oils to biodiesel through transesterification. This is because the transesterification process involves removing the glycerides and combining oil esters of vegetable oil with alcohol, and in doing this viscosity reduces to a value that is compatible with diesel.

- Vegetable oil has significantly dissimilar injection, atomization, and combustion characteristics to diesel fuel

- Vegetable oil has high viscosity, which interferes with the injection process and leads to poor fuel atomization, low volatility both of which cause bad combustion in CI engines

- Vegetable oil, when inefficiently mixed with air, can contribute to incomplete combustion and therefore heavy smoke emissions

- Vegetable oil has high cloud and pour points, which can cause problem during cold weather

- Further problems incurred by using vegetable oil are lube oil dilution, high carbon deposits, ring sticking, scuffing of the engine liner, and injection nozzle failure

Despite the promise of an alternative fuel, biodiesel does claim some drawbacks. A full record of these disadvantages, created through literature research, can be seen in

Disadvantages	Explanation
Biodiesel from biomass has a several ecological issues	Biomass creation is land intensive and requires rich agricultural land for growth of feedstock. This limits ground for growing food [3]. Biomass requires pesticides and fertilizer which can pollute nearby water resources [3].
Biodiesel is expensive [3, 7, 9]	Biodiesel is dependent on the production of vegetable oil which is a lot less making biodiesel more expensive [7, 9]
Biodiesels chemical properties are affected by cold weather	Biodiesel has weak cold flow properties [3] due to the increase in its density at low temperatures [7, 9]. Also the higher surface tension and viscosity deteriorate the cold start stability [9].
Biodiesel as a fuel may have higher NOx emissions [9]	Biodiesel running in an engine has a negative effect on the injection system (injector coking, fuel lines clogging, etc.), on combustion (poor atomization, carbon deposits, etc.) and on the hardware (piston ring sticking). Due to Biodiesels effect on ignition delay and some of the above listed effects, the vaporization process is negatively inclined, so that incomplete combustion is induced during the engine start and also an increase of NOx emissions can be noticed [9].

Table 4.
A table showing the disadvantages of biodiesel in literature.

Table 4. To overcome the many failings of biodiesel, there is the option of blending the fuels with petroleum diesel. By doing this, there is an opportunity to use the fuel in sub-freezing conditions [7, 9]. Another way to improve its low temperature properties, which is currently being researched, is by applying cold filter clogging [37]. Further reasons for blending include significantly improving GHG emissions and increasing engine efficiency [14].

3. Key chemical and physical properties of biofuels

The performance, combustion, and emission characteristics of fuels depend on their properties. Many researchers state that the most important characteristics for fuel application are density, viscosity, lubricity, heating value, cetane number, flash and fire points, cloud and pour points due to their strong control over emissions characteristics and engine efficiency, ultimately indicating the quality of the fuel [3]. Before using any fuel in a compression ignition engine, its properties must measure as specified by standards. This section presents the main fuel properties that affect the quality of fuel [38–40].

3.1 Flash and fire point

The flash point is the temperature at which the liquid vapors ignite under normal pressure. The flash point is measured according to ASTM-D93 [41].

3.2 Density

Density is the mass per unit volume, and for diesel, fuel is normally determined at 15°C [21]. The composition of the fuel controls the density as diesel/biodiesel fuels are made up of a wide range of different hydrocarbon compounds of various densities and molecular weights. Density affects engine power, emissions, and fuel consumption particularly in contact injection systems [21]. There are strong correlations between viscosity, cetane number, and aromatics content, higher density increases energy concentration of fuel and minimizes fuel leakage, as well as encouraging the fuel optimization efficiency [4]. However, higher density does also incur higher viscosity, which consequently decreases engine performance and worsens emissions. For these reasons, the density must be meticulously regulated to be within a relatively narrow range [4, 21].

3.3 Lubricity

Lubricity has come to be a vital quality, mainly for biodiesel, because of better pressures in new diesel gas injection (DFI) generation, which needs higher lubrication from the gas [5, 38]. Lubricity is surely the capacity of the gas to lessen friction among shifting elements inside the engine to assist it run smoothly. Up to 30% of mechanical strength is fed on through friction, making it key to attaining the goal of growing engine performance through decreasing losses [42, 43]. A "dry gas" is a gas that is not made of a good enough quantity of lubricating components inflicting it to be not able to lubricate additives such as gas transport and injection system, cylinder liners, etc. Tribological assessments may be executed to evaluate lubricity, and on this record High-Frequency Reciprocating Rig (HFRR) system might be used as a different technique, which includes Scuffing Load Ball-on-Cylinder Lubricity Evaluator (SLBOCLE) having now no longer been used a lot in view in 2005 and a literature assessment of consequences is desired [44]. Diesel refiners now no longer impart a lubricity price in any respect of their specs for wholesalers, and the reality that the ASTM recommendations best installed a fashionable lubricity requirement in 2005 [45]. Supposedly, as a lot as 30% of mechanical strength is fed on through friction [23, 42, 43]. This makes gas lubrication essential to lessen scarring of additives within the engine. Biodiesel gas blends provide considerably better lubricity than traditional diesel, for instance, B2 can offer as much as a 65% development in lubricity [46].

3.4 Cloud and pour point

The pour point of a liquid is the lowest temperature at which liquid can flow when it is cooled. The pour point is measured according to ASTM-D97. The cloud point is the temperature at which the solid crystals or wax is formed in cloudy color when it is cooled. The cloud point is measured according to ASTM-D2500 [41].

3.5 Viscosity

Viscosity is a measure of "thickness" of a fluid and its resistance to deformation. Kinematic viscosity is the ratio of dynamic viscosity to density. Viscosity is dependent on fuel composition and so is replicated in the distillation parameters, density, and cold flow properties. A low viscosity is preferred as a very high viscosity may cause fuel pump distortion, which encourages the production of engine deposits and delays combustion [10]. In addition, there is a proven negative correlation between increasing temperature and the viscosity of a fuel [11]. It is the most important property of fuel when aiming to preserve the performance of a diesel operation engine [11]. A higher viscosity means the greater the affinity of the fuel to induce the formation of engine deposits [38], incurring poor fuel spray [3], atomization, and also insufficient fuel flow. It also delays combustion [11]. Furthermore, a very high viscosity may cause fuel pump distortion [21]. In these respects, lower viscosity is desired [3].

3.6 Heating value, gross calorific value

A fuel's heating value, calorific value of combustion are a gauge of the quantity of thermal strength it releases for the duration of its burning and is an influential component within the gasoline economic system and energy deliverability [5, 22]. There are several values to consider when concentration on the available energy within a fuel, the lower heating value (LHV) and the higher heating value (HHV) included. These differ, as the lower HV does not include energy in the combustion of water vapor, whereas the higher does. It is known that higher moisture contents in the fuel have a negative correlation with both LHV and HHV as the water takes up volume, which could have been fuel and must vaporized before the fuel can ignite. The net heating value is the appropriate quantity for comparing fuels as the engine exhaust water in gas phase [21]. A diesel engine fuel is of better calorific value as it allows the heat launch for the duration of combustion and improves engine's overall performance for the duration of combustion [4, 11]. The three fundamental elements that influence automobile gasoline economic system, torque, and horsepower are the performance of the engine turning strength within the gasoline into usable work, kind of engine (fuel or diesel), and the gasoline's volumetric strength content material or heating value [5, 21, 23].

3.7 Cetane number

In terms of Spark Ignition and Compression ignition engine types, each requires opposite abilities with regard to speed of combustion. For example, gasoline (SI) engine prefers a high octane rating as it will then be able to withstand more compression before igniting. The opposing diesel (CI) engine favors low octane ratings, as fuel is injected once air has already been compressed. The cetane number is the inverse of octane number and therefore is defined as ease of combustion in a compression setting so a lower one is required for SI and higher ratings for CI engines. The cetane number is also a prime indicator of the quality of fuel used [4] and directly affects its combustion quality [11]. As there is a negative correlation between cetane number and ignition delay time, a shorter delay causes a higher cetane rating [3, 4, 11, 21], which is desired by a CI engine for smooth operation, with nice cold start behavior at low noise, and additionally provides greater fuel economy and power [3, 7, 21].

4. Properties of crude vegetable oil

Research was gathered on the above key physicochemical properties for vegetable oils before they have been altered to biodiesel. The data can be seen in **Table 5** below along with the average values for diesel gathered from several sources.

The density is presented in **Table 5** and shows the variety of oils to not be far from diesel, with percentage differences of all the oils being in the range of about 8.04–13.23% above diesel. However as stated earlier, there is narrows gap acceptable for density, so engine testing would have to be done to see if this was suitable and also as previously specified, a high density will lead to a high viscosity, which is not desired. This can be seen to be true in **Table 5**, as it can be seen that all the unprocessed vegetable oils have a very high viscosity with the maximum gathered being for crambe oil, which is almost 15 times higher than that of diesel. The minimum viscosity of vegetable oils is for linseed, which is still seven and a half times as much a diesel. The large molecular mass of these oils is what causes the elated viscosity. The molecular weight of the oils is in the range of 600–900, which is about 20 times higher than that of diesel fuel [9]. High viscosity in a fuel can cause fuel flow and ignition problems in unmodified CI engines and also decreases in power output [48].

The best and highest calorific value, in terms of mass, which can be seen in **Table 5**, is for diesel. The vegetable oils are between 9 and 12% lower than its value of 44.651 MJ/kg due to the presence of chemically bound oxygen, which lowers their heating values by this percent [9]. The best vegetable oils are beef tallow and crambe, which both have a heating value of 40.5 MJ/kg. However, the fuel with the most superior cetane number is not diesel, but palm oil has a value of 52 compared to diesel's 50.1. All other vegetable oils are inferior to diesel in this regard, and linseed has the lowest number, which is 15.5 less than diesel.

Another interesting calorific value for fuels is MJ/l as this factor is in density. It is worked out by multiplying the calorific value (MJ/kg) by density (kg/l). These were determined from the data gathered from literature and are presented in the fourth

Fuel	Density (kg/m^3)	Calorific value (MJ/kg)	Kinematic viscosity at 40°C (mm^2/s)	Lubricity (μm)	Cetane number	References
Canola	910	39.78	37.7	—	41.5	[3]
Palm	925	39.3	38.25	—	52	[3]
Beef tallow	920	40.5	NA	—	NA	[3]
Corn	909.5	39.5	34.9	—	37.6	[9]
Cotton seed	914.8	39.5	33.5	—	41.8	[9]
Crambe	904.8	40.5	53.6	—	44.6	[9]
Linseed	923.6	39.3	27.2	—	34.6	[9]
Peanut	902.6	39.8	39.6	—	41.8	[9]
Rapeseed	911.5	39.7	37	—	37.6	[9]
Safflower	914.4	39.5	31.3	—	41.3	[9]
Sesame	913.3	39.3	35.5	—	40.2	[9]
Soya bean	913.8	39.6	32.6	—	37.9	[9]
Sunflower	916.1	39.6	33.9	—	37.1	[9]
Palm	918	NA	39.6	—	42	[9]
Babassu	946	NA	30.3	—	38	[9]
Karanja	NA	34	27.84	—	NA	[9]
Diesel	800-845	35	1.9-6	300	51	[3, 4, 8, 9, 11, 21, 22, 42, 47]

Table 5.
Important physiochemical properties of vegetable oils and diesel.

column of **Table 5**. It can be seen that the calorific values shift significantly closer to diesel. This is represented by the range going from 9 to 12% lower, in kilograms terms, to a range of just 0.04–3.1% lower, in liter terms, than that of diesel. This is a significant drop in percentage error. Diesel's calorific value actually drops by incorporating the density, though it still claims the highest value in the table. In terms of aiding combustion by increasing heat release and therefore performance, diesel is the best fuel shown in the table. Linseed's shorter ignition delay makes it valuable, even though it fails in all respects. Further processing is required to improve the properties of vegetable oil and bring them closer to that of diesel so they can be immediately implemented in CI engines. The transesterification process produces more favorable physicochemical parameter fuels, which have wider applications than their corresponding vegetable oil counterparts [49]. The methyl esters of Canola, Palm, and Beef Tallow shown in **Table 5** are examined in the following section.

5. Properties of biodiesel

Literature was investigated to find key physicochemical properties for biodiesel and synthetic diesel produced by either transesterification or Fischer-Tropsch Synthesis. The data can be seen in **Table 6** below along with the average values for diesel. All transesterification-produced biodiesels in the table are methyl esters.

Firstly by looking at the first three rows of data in **Table 6**, which were taken from McCarthy et al. [48] and Arbab et al. [4] for canola, palm, and tallow, it can be processed that the goal of transesterification to reduce the viscosity of vegetable oil is

Fuel	Density (kg/m^3)	Kinematic Viscosity at 40°C (mm^2/s)	Calorific Value (MJ/Kg)	Lubricity (μm)	Cetane number	References
Diesel Standard Min/Max	800–845	1.9–6	35	300	51	[3, 4, 8, 9, 11, 21, 22]
Canola—(methyl ester)	887.5	4	40.07	—	41.5	[3]
Palm oil—(methyl ester)	866.75	4.6175	40.3225	—	58.5	[3, 4]
Tallow—(methyl ester)	877	4.6	39.9	184.5	58	[3, 42]
Jatropha	852	4.3233333	40.8333	—	50.5	[3, 4, 11, 22]
Coconut	887	3.355	36.55	—	55.5	[3, 4]
Cotton seed	879.5	4.45	40.64	—	51.725	[3, 4, 11]
Sunflower	879.5	5.2	40.13	—	50.5	[3, 4]
Soybeen	890	4.525	39.035	193	46.5	[3, 4, 42]
Rapeseed	861.5	4.35	38.525	205	50.95	[3, 4, 50]
Marula oil	813	3.74	42.2	—	—	[22]
Marma bean (Tylosema esculentum)	846	3.5	42	—	—	[22]

Fuel	Density (kg/m³)	Kinematic Viscosity at 40°C (mm²/s)	Calorific Value (MJ/Kg)	Lubricity (μm)	Cetane number	References
Mongongo (Manketti oil)	817	3.86	41.7	—	—	[22]
Karanja	883	8.2925	36.56	—	55	[9, 11]
Polanga	899.3	4.67	40.275	—	57.3	[11]
Mohu	910	4.89	39.655	—	51.5	[11]
Rubber seed oil	870.5	5.885	38.735	—	43	[11]
Jojoba oil	864.5	22.3	45.017	—	63.5	[11]
Tobacco oil	874.25	3.865	39.12	—	50.30	[11]
Neem	938.5	34.5	36.6	—	51	[11]
Linseed oil	907.5	26.4	38.35	—	31.5	[11]
FT diesel	773.5	2.7925	45.505	211	77	[5, 13, 21, 47]
Diesel	800-845	1.9-6	35	300	51	[3, 4, 8, 9, 11, 21, 22, 42, 47]

Table 6.
Important physiochemical properties of biodiesels and diesel.

achieved. The values for both palm and canola are reduced by about 33.7 mm²/s. This makes the biodiesels more suitable for CI engines. The three fuels all have reduced density than that of their vegetable oil counterparts with improvements between about 2.5 and 6.3%, but still have higher density than diesel, which would result in higher fuel consumption [48]. The calorific values of canola and palm also see a slight increase where tallow sees a minor decrease, though they still do not obtain the value of 44.651 MJ/kg for diesel. This means that the biodiesel may incur high fuel consumption and less power in a diesel engine, with tallow performing the worst. The power reduction can be between 5% and 10% depending on the biodiesel, engine speed, and load [48]. The cetane numbers of palm and tallow, however, rise well above diesel by 16.8% and 15.8%, respectively. It would be true to say for the three mentioned methyl esters that palm is the better source, as if used as engine fuel it would be more efficient. Among the various types of oils, there are differentiations in the types of fatty acids in their chain making them either, saturated, monounsaturated, or polyunsaturated. Biodiesel properties can be affected by these ranges of chain saturation levels [48].

5.1 Density

Overall, from **Table 6** and **Figure 3**, it can be said that the biodiesels usually have a higher density than diesel, with only SB100, SR100, and FT diesel failing outside of that rule. However, it also can be seen that Jatropha, SB100, SR100, and TE100 have the closest density to diesel all coming in under 3% difference. Neem biodiesel has the highest density with an average of 938.5 kg/m³ compared to diesels at 835.126/m³. Then Fischer-Tropsch diesel has the lowest density of all the fuels at 773.5 kg/m³, which is a 7.076% variation from diesel. When looking at B100 fuel, pure biodiesels, and

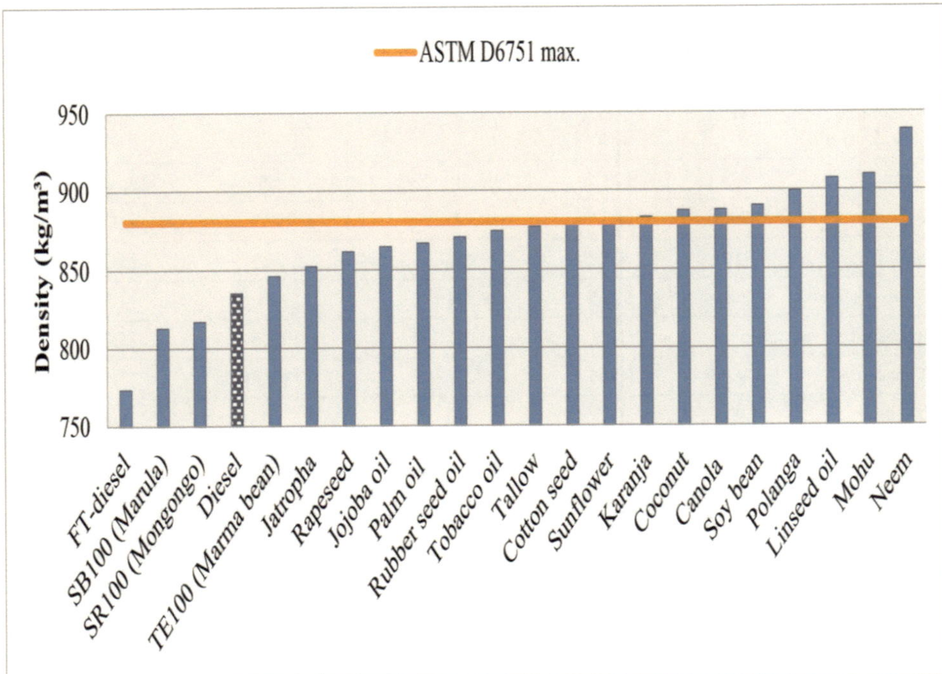

Figure 3.
Density comparison of various kinds of biodiesels.

petroleum diesel, ASTM D6751 standards must be followed. According to Ali et al. [8], the density for these pure fuels should not exceed 880 kg/m^3, a graph is presented in **Figure 3**, which shows the density of the evaluated fuels with the ASTM limit visible. Only FT-diesel, SB100 (Marula), SR100 (Mongongo), TE100 (Marma bean), Jatropha, Rapeseed, Jojoba oil, Palm oil, Rubber seed oil, Tobacco oil, Tallow, Cotton seed, and Sunflower remain within this specification making them the most suitable fuels for use in this instance.

5.2 Viscosity

In **Table 6**, the examined biofuels can be seen to have viscosities ranging from 2.79 to 34.5 mm^2/s, with FT diesel at the bottom end and Neem biodiesel at the top end of this range. It is stated in ASTM D6751 standards. For B100 and diesel fuels, that viscosity should not be below 1.9 or above 6 mm^2/s. **Figure 4** contains a graph with the viscosities of the fuels except Jojoba, Linseed, and Neem biodiesel, which are incredibly vicious, and compares them to the maximum value of ASTM D6751 limitations, as non-fall below 1.9 mm^2/s minimum. It can be seen that Karanja biodiesel also does not meet the specification requirements and exceeds the limit by 2.2925 mm^2/s. All other biodiesels as shown in t **Figure 4** are acceptable and could be effectively used in diesel engines since they largely satisfy the fluidity requirements of alternative biodiesel fuel [20]. However, with the aim of a low-viscosity FT, diesel proves to be the best fuel, though coconut and TE100 also improve on diesel with lower values and therefore would give enhanced combustion and atomization [4, 11].

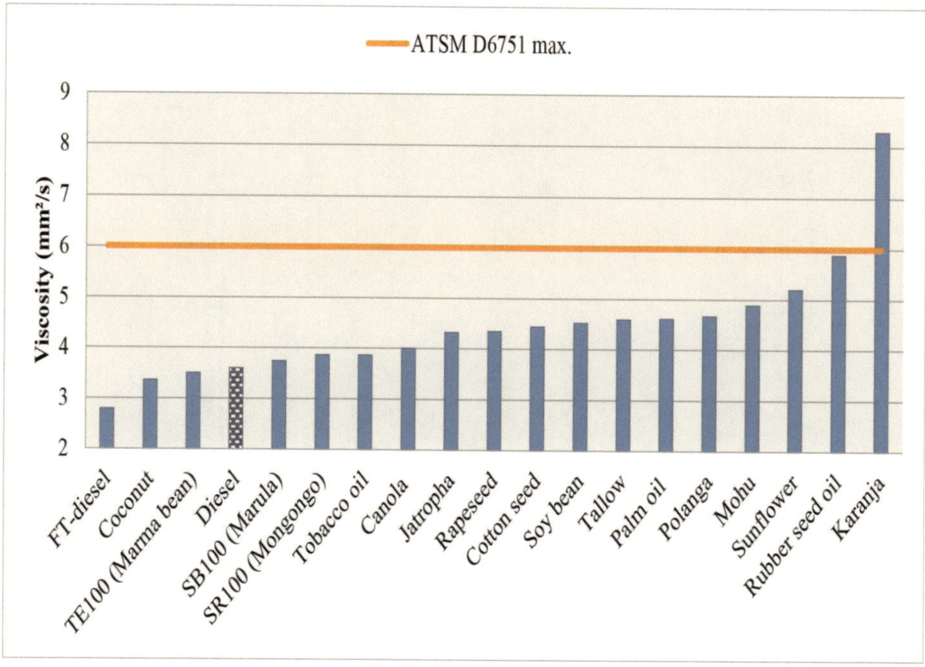

Figure 4.
Viscosity comparison of various kinds of biodiesels.

5.3 Lubricity

Results from literature for this test can been seen in **Table 6** for several alternative diesel fuels. The smaller the wear scare, the better the lubricating properties of the fuel, and they must comply with ASTM maximum value of 300 μm. It can be seen that soybean 20% blend biodiesel forms the smallest wear scar and diesel performs the worst; however, the values of diesel found in text fluctuated a lot so it is difficult to make a comparison here. Therefore, it would be suggested that a control diesel lubricity test should be done every time. Unfortunately, no Biomass-To-Liquid (BTL) Fischer-Tropsch diesel could be found, so Gas-To-Liquid (GTL) from Natural Gas is compared here. It is also suggested that some BTL FT fuel be experimented on for a better comparison. Overall, the biodiesels are seen to improve the lubricity of diesel from **Figure 5**.

5.4 Calorific value

The calorific value of all the biofuels in **Table 6** can be seen to be less than petro-diesel with the exception of Jojoba oil and FT diesel, this is due to its higher oxygen content [11]. As was stated earlier, a high heating value is desired, though a minimum value is not specified in the biodiesel standards ASTM D6751 but is prescribed in EN 14213 at 35 MJ/kg [8]. The biofuels are judged against this standard in **Figure 6** below, and it can be seen that all of the fuels have satisfactory heating values. Coconut-derived biodiesel is the closest to the EN14214 value at 36.55 MJ/kg, which is an 18% decrease from petro-diesel in its calorific value, while FT diesel achieves 1.91%

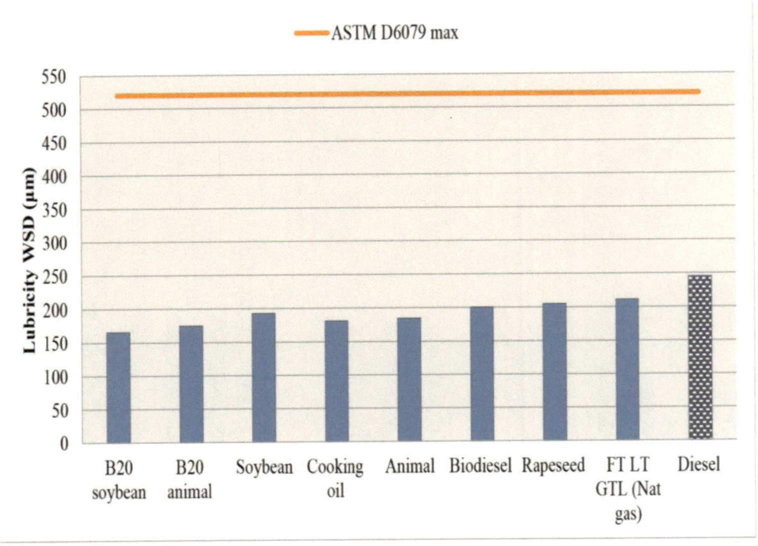

Figure 5.
Lubricity comparison of various kinds of biodiesel blends.

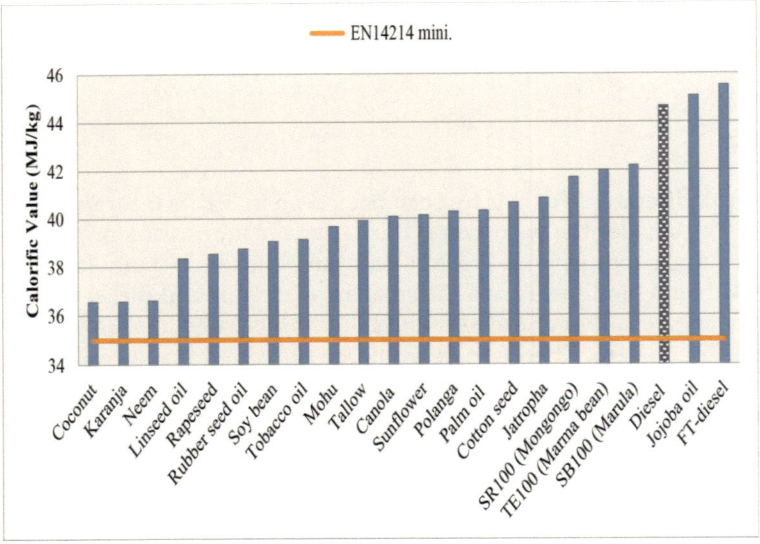

Figure 6.
Calorific value comparison of various kinds of biodiesels.

increase with Jojoba close behind at 0.94% increase. This shows that Jojoba biodiesel and FT diesel would give better engine performance and are attractive diesel substitutes [11, 21]. Though there is variation in the values as is exemplified in **Figure 6**, if it was stated that fuels within 10% of petro-diesels calorific value were comparable, this would encompass Palm oil methyl etser, Jatropha, Cotton seed, SB100 (Marula), TE100 (Marma bean), SR100 (Mongongo), FT diesel, Polanga, and Jojoba oil giving a wide range of alternative fuel options.

5.5 Cetane number

CN of a fuel is a measure of its propensity for auto-ignition. Cetane number has a strong impact on the time interval between the fuel injection and the combustion in the diesel engine. The majority of fuels had a higher cetane number than petro-diesel though Linseed, Canola ethyl ester, Rubber seed, and Soybean do fall below diesel and in fact below the ASTM minimum requirement at 47, making them unsuitable as diesel fuels. A comparison of the cetane number and the specification necessity can be seen in **Figure 7**. FT diesel and Jojoba oil biodiesel are again the fuels with the best and highest values, 77 and 63.5, respectively. This is an increase of 53.69% and 26.75% on petro-diesel, and in this respect these two alternative diesels are better than others, which indicates better auto-ignition quality [21]. The reason the cetane number of biodiesel is usually higher than that of petro-diesel is its longer fatty acid carbon chains and the presence of saturation in molecules. The cetane measurement is dependent on two compounds, namely hexadecane and heptamethyl nonane [11]. Though the fuels with the best cetane number have been highlighted, it should be noted that any of the fuels, which follow the guidelines, are acceptable for use in CI engines.

6. Properties of biodiesel blends

When implementing biomass-derived fuels in modern engines, failure can occur due to high viscosity, density, and deposit build-up. A solution with potential is to blend the mixtures with fossil fuels [35]. This will also aim to solve the fundamental problem associated with growing plants specifically for energy, the trade-off between foods and fuels [51]. Biodiesel blends between are represented in such a way that the number value signifies the percentage of biodiesel contained in the blend; for

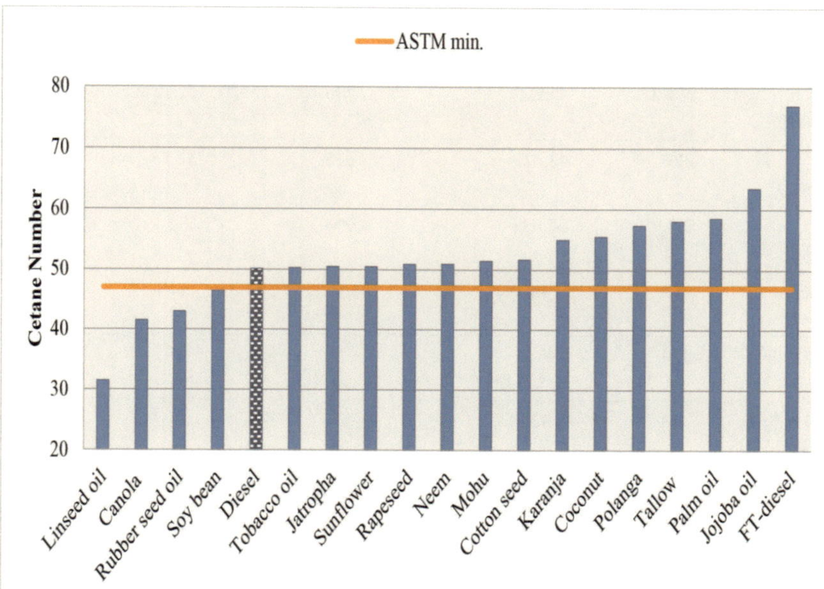

Figure 7.
Cetane number comparison of various kinds of biodiesels.

example, B30 contains 30% biodiesel and 70% petro-diesel. B6 to B20 should follow the specification for ASTM D7467, and the data from several sources for biodiesel blends are presented in **Table 7** below with average values for diesel, gathered from several sources [8, 9, 13]. All of the biodiesels used in the blends are methyl esters.

Fuel	Density (kg/m^3)	Calorific value (MJ/kg)	Kinemtaic viscosity at 40°C (mm^2/s)	Cetane number	References
B10 (palm oil)	850	44.23	3.86	NA	[8]
B20 (palm oil)	853	44.12	3.91	NA	[8]
B30 (palm oil)	857	43.13	3.95	NA	[8]
B40 (palm oil)	860	42.95	3.97	NA	[8]
B50 (palm oil)	863	42.74	4	NA	[8]
B100 (palm oil)	880	38.57	4.61	NA	[8]
B20 (Karanja)	NA	38.28	3.39	NA	[9]
B40 (Karanja)	NA	37.85	4.63	NA	[9]
B60 (Karanja)	NA	37.25	5.42	NA	[9]
B80 (Karanja)	NA	36.47	6.56	NA	[9]
B100 (Karanja)	NA	36.12	9.6	NA	[9]
SB30 (soya bean)	855.4	43.619	3.419	51.7	[13]
SB50 (soya bean)	866.9	42.547	3.571	52.1	[13]
SB80 (soya bean)	874.1	40.705	3.958	59.4	[13]
RB30 (rapeseed)	854.2	43.658	3.496	52.1	[13]
RB50 (rapeseed)	865	42.412	3.636	52.7	[13]
RB80 (rapeseed)	873.9	40.698	4.094	60	[13]
B5 (Citrullus Lonatus)	NA	44	4.9	54	-
B10 (Citrullus Lonatus)	NA	43.2	4.5	55	-
B15 (Citrullus Lonatus)	NA	43	5	56	-
B20 (Citrullus Lonatus)	NA	42.3	5.2	57	-
Diesel	800-845	35	1.9-6	51	[3, 4, 8, 9, 11, 21, 22, 42, 47]

Table 7.
The important physiochemical properties of biodiesel blends and diesel.

6.1 Density

The density values that can be seen in **Table 7** show that B10 (palm oil) has the lowest density at 850 kg/m^3 for the biodiesel blends, but in fact all of the blends have a higher density than petro-diesel at 835.126 kg/m^3. This is a similar outcome to biodiesels by themselves, as the majority of them had higher densities than diesel. It can be seen in **Figure 8**, which represents the blends' densities and compares them to ASTM specifications, that blending in more diesel fuel into the biodiesel lowers the density, as expected. The range of variation from diesels density is 1.78%–5.37%, so relatively there is very little differentiation. In addition, it should be noted that B10 and B20 for palm oil fall under the ASTM D7647 limits that control the B6-B20 blends making them viable options. The density of POME (Palm Oil Methyl Ester) is the highest; accordingly, the density of the blended fuel B30 is 2.6% lower than that of B100 due to the effect of blending with diesel [8]. Furthermore, at biodiesel blending ratios of more than 30% (B30), it can be noticed that they do not comply with ASTM D7647, as their densities are too high; however, they are not covered by this specification. They do comply with ASTM D6751 for pre-blended biodiesels.

6.2 Viscosity

The viscosity results for bended fuel samples range from 3.39 mm^2/s for B20 (karanja) to 9.6 mm^2/s for B100 (karanja) as can be seen in **Table 7**. These data are also visible in **Figure 9** below excluding B100 (karanja), which is extremely viscous, almost 2.67 times that of petro-diesel, and way beyond the limitation of both ASTM D7467 and ASTM D6751. Accordingly, the viscosity of the blend decreases as the methyl ester amounts decrease in the fuel mixture. However, the blended fuel viscosity still meets the blended fuel standard requirements ASTM D7467 for up to 50% biodiesel blending ratio (B50) and RB80. Between B40-B80 Karanja and B100 POME are within the standard requirements ASTM D6751 for pre-blended biodiesel.

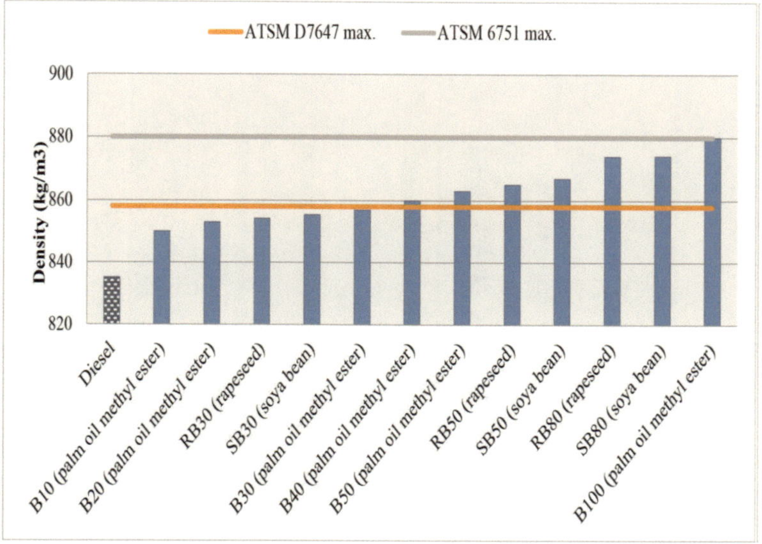

Figure 8.
Density comparison of various kinds of biodiesel blends.

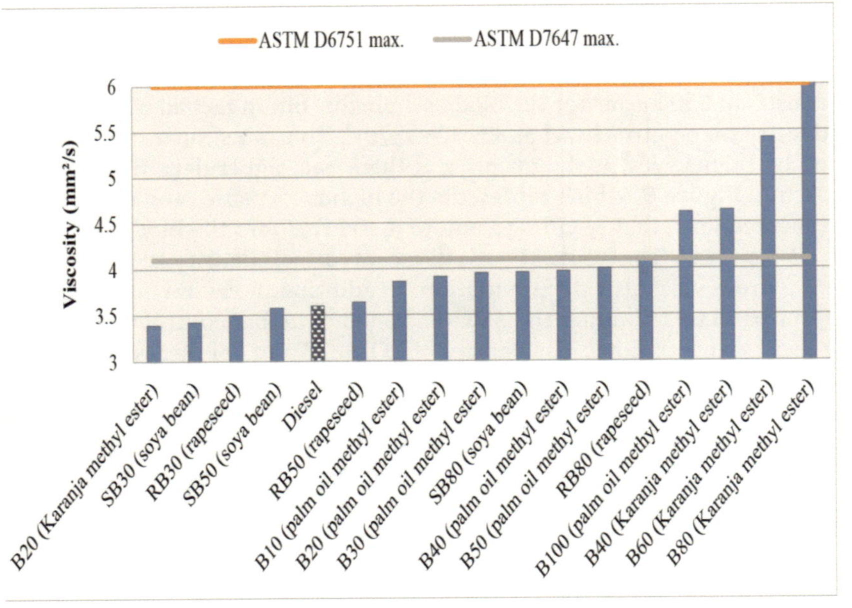

Figure 9.
Viscosity comparison of various kinds of biodiesel blends.

6.3 Calorific value

As mentioned in the previous section, the heating value is not specified in the biodiesel standards ASTM D6751 but is prescribed in EN 14213 with a minimum value of 35 MJ/kg [8]. The values of heating value from **Table 7** are compared to this limitation in **Figure 10,** and all of them obtain and exceed this minimum requirement.

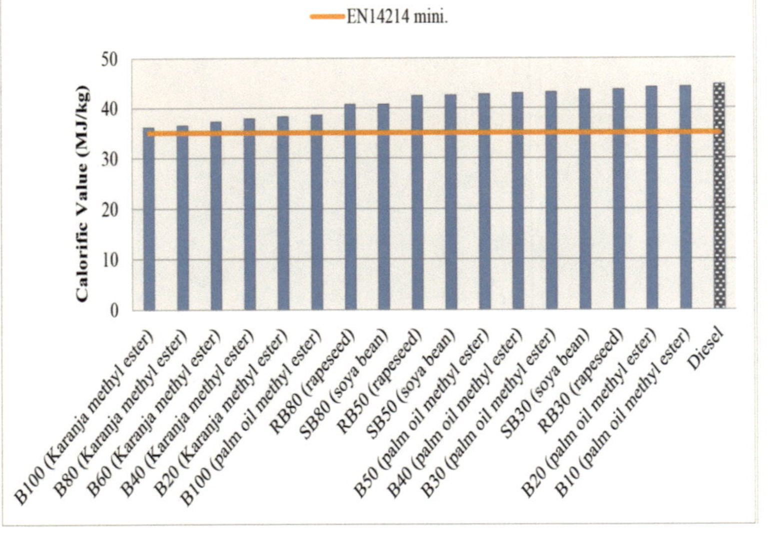

Figure 10.
Calorific value comparison of various kinds of biodiesel blends.

It can be seen in **Figure 10** that the heating value of the POME-diesel blend decreased with a higher volumetric percentage of the POME, the same could be said for Karanja, Soya bean, and Rapeseed diesel blends. The range of differentiation from Petro-diesels calorific value that can be seen is 0.94–19.11% with B100 Karanja at the bottom end and B10 POME with the closest value to diesel. Accordingly, the blended fuel heating value decreases with increasing biodiesel ratios in blended fuel, which is attributed to the relative composition of biodiesel, which has oxygen, present in the structure. Oxygen is not a component of the conventional diesel and results in reduced carbon and hydrogen contents [8].

6.4 Lubricity

Biodiesel gasoline blends provide extensively higher lubricity than traditional diesel; for instance, B2 can offer as much as a 65% development in lubricity [46]. To assess lubricity, a tribological check may be carried out, which measures the scale of the wear and tear mark in an HFRR (excessive frequency reciprocating rig) check [42]. Results from literature for this check can been seen in **Table 7** for numerous opportunity diesel fuels [42, 43, 47, 50, 52–54]. The smaller the wear and tear scare, the higher the lubricating homes of the gasoline, and they need to follow ASTM D6079 most cost of 520 μm [5, 23].

7. Engine performance with biodiesel

This section is going to concentrate on biodiesel as a whole as well as Fischer-Tropsch diesel. First, the engine performance will be examined by combining data from several different reports. **Table 8** below shows biodiesel and F-T diesel

Fuel-biodiesel and its blends	References	CO	CO_2	NO_x	HC	Smoke
Mahua oil; B20, B40, B60, B100	[55]	Decrease	Increase	Increase 11.6%	Decrease 32%	—
Karanja oil; B20, B40, B60, B80, B100	[55]	Decrease 94% and 73% for B20 and B100	—	Decrease 26%	Increase	Decrease 80% and 20% for B20 and B100
Jatropha oil; B10, B20, B40, B50 and B100	[55]	Decrease 24% for B100	—	Increase 24% for B100	Decrease	Decrease 45% for B70 and 60% for B100
Cottonseed oil; B10, B20, B30, B40, B50, and B100	[55]	Decrease 24% for B30	Increase	Increase 10% for B30,	Decrease	Decrease 14% for B10
Soybean oil; B100	[55]	Decrease	Increase	Increase	Decrease	Decrease
Rapeseed oil; B5, B20, B70 and B100	[55]	Decrease 12%, 25%, 31% and 35%	—	Increase 12% for B100, 9% for B70 and 6% for B20	Decrease	Decrease 45% for B70 and 60% for B100

Fuel-biodiesel and its blends	References	CO	CO_2	NO_x	HC	Smoke
Jojoba oil, B100	[55]	Increase	—	Increase	Increase	—
Neem oil; B5, B10, B15	[55]	Decrease	—	Increase	Decrease	Decrease
Waste cooking oil; B5, B10, B20 and B30	[55]	Decrease 6.75%, 7.33%, 8.32% and 13.1%	—	—	Decrease 105%, 19.9%, 27.7% and 36%	—
Aprocpt seed kemel oil; B5, B20, B50, B100	[56]	Decrease	Decrease	Increase	Decrease 18.66% for B100 and 2.66% for B5	Decrease
Rice bran oil, B100	[55]	Decrease 25.8%	Increase 20%	Increase 4%	Decrease 54%	Increase 27.93%
Castor Bean oil; B10, B25, B50 and B100	[55]	Decrease	Decrease	Increase 44.6%	—	—
Paradise oil, B20, B40, B50 and B100	[55]	—	—	Increase 5% and 8% for B50 and B100	Decrease 22% and 27% for B50 and B100	Decrease 33.5%, 39.4% for B50, B100

Table 8.
Comparison of diesel and different biofuels' emissions at different test conditions.

compared to petro-diesel in term of engine performance from two papers, which had gathered information from several sources.

7.1 Power

Engine power is an important characteristic that affects vehicle acceleration and handling. Using **Table 8** above it can be said that most researchers agree that power and engine torque for biodiesel decrease, with 70.4% majority. This is generally established to be due to the lower heating value and high viscosity of biodiesels, though some fluctuations in results can be noted [3, 4]. FT diesel, on the other hand, sees a unanimous performance between similar and decreased power display. In regard to biodiesel blends, it would be true and as expected to say engine power increases with a decreasing biodiesel ratio, due to the increasing amount of petro-diesel in the mixture. As a comparison, according to Ali et al., the brake power achieved for diesel fuel was about 0.5, 1.6, and 2.7% higher than that of the blended fuel B10, B20, and B30, respectively, at the same engine conditions [8]. Also Murugesan et al. [9] reported that the torque produced for B20 and B40 was 0.1–1.3% higher than that of diesel due to complete combustion of fuel.

7.2 Economy

The Brake-Specific Fuel Consumption (BSFC) is a gauge of the fuel flow rate per unit power output and is a measure of an engine economy performance [8]. In **Table 8** it can be seen that pure biodiesel was found to more commonly increase the fuel consumption of an engine when compared to diesel by agreement between 87.1% of the references studied, whereas FT diesel sees a decrease with a settlement of 83%. Biodiesel has a Higher BSFC due to its lower calorific value [3, 4, 7, 8, 11, 13], though some researchers have demonstrated that the increase of biodiesel fuel consumption ratio is more than lower heating value ratio. With this in mind, they determined that the high density of biodiesel could also contribute to this factor [3]. However, biodiesel blends with 20% or less biodiesel content have been noticed to reduce fuel consumption and give higher brake power because of complete combustion [4]. As a comparison, the BSFC of B20 and B40 was found to be 0.8–7.4% lower than diesel in one case [9]. Looking back at **Figure 6** in the previous section, it is displayed that the only two diesel fuels with a higher heating value than petro-diesel are FT and jojoba biodiesel, which explains the decrease in fuel consumption displayed in **Table 8** for FT diesel and is why jojoba would be expected to also show a decrease. This could be one of the sources of the inconsistent trends in **Table 6** for biodiesel with 12.9% of the sources finding the BSFC to be similar or lower than that of diesel.

7.3 Efficiency

The Brake Thermal Efficiency (BTE) is the ratio of the thermal power available in the fuel to the power that the engine delivers to the crankshaft [8]. However, in one case where the efficiency stayed the same, despite the better fuel consumption of FT diesel, it was noted that parameters such as injection pressure, injection timing, and EGR rate among others were more important in maximizing the engine efficiency than the calorific value of the fuel [13]. This could explain the fluctuations in the data provided in **Table 8**. For biodiesel blends it was found in some instances that the BTE slightly increased with increasing biodiesel ratios due to the high oxygen content of the blended fuel compared to mineral diesel, which enhances the fuel combustion process, and the additional lubricity provided by the biodiesel [8]. In other cases, this increase in efficiency was found to be true up to a point, say B40, but then from B60 to B100 efficiency decreased and became lower than that of petro-diesel [9]. Density affects engine power, emissions, and fuel consumption particularly in constant injection systems [21]. Many research studies have been done to investigate the effect of using biodiesel on engine performance and efficiency for examples: Use of kernel oil as biodiesel in four-stroke diesel engine increases brake specific fuel consumption and reduces the brake thermal efficiency [57] due to the lower calorific value and higher viscosity of the biodiesel [56]. In addition, Jatropha biodiesel, linseed biodiesel, karanja oil methyl ester, and neat vegetable oil decrease brake thermal efficiency [55, 56]. Rao et al. [58] stated that decrease in thermal efficiency is due to the early start of combustion, increase of compression work and heat loss. In contrast, some experimental results stated that increase in efficiency of engine by using biodiesels, for example, laforgia, observed that use of biodiesel in an indirect injection diesel engine increases the thermal efficiency by 10% with comparison of diesel fuel [59]. By investigation of all researchers' works, it is clearly observed that the efficiency in high load increases in all types of fuels including diesel fuel. This increase in use of biodiesel is lower than use of diesel in CI engine. Mahua, palm, and jatropha biodiesels decrease the efficiency higher than other types of biodiesels.

The performance of an engine is reflected in power output, fuel consumption, efficiency, and economy. Nearly all sources agree that alternative fuels, such as biodiesel and FT diesel, can be used in diesel engines with little or no modifications [60].

To improve the properties of biodiesel blending is an option, either with petro-diesel or by blending biodiesels from two or more feedstock [3].

7.4 Emissions characteristics of biodiesel

All engines produce exhaust gases when they run. These gases, however, are neither good for the environment nor human health. Use of fossil fuels in transportation facilities increased emissions of harmful pollutants. The hydrocarbons from engine exhausts form ground-level ozone, which is the major element of smog. Ozone affects human health such as lung disease and eye irritation and many other cancer issues [55]. Due to these reasons, fuels must meet the stringent emission norms set by the different regulating authorities throughout the world. The main pollutants from engine exhaust gases, which are considered for this review investigation, are carbon monoxide, carbon dioxide, nitrogen oxides, hydrocarbon, and particulate matter.

Table 8 presents the emission characterization from several research works. Results were shown by increase and decrease with the help of readers to understand the advantage and disadvantage of biofuels in the case of emission. Karanja oil blends decrease the CO by 94% and smoke by 80%, rapeseed oil blends decrease the CO by 35%, and rice bran oil blends decrease HC by 54%. Most of the biofuels increase the amount of NO_x in exhaust emissions, which can use some after treatment to solve this issue. In conclusion, types of biodiesels, purity, types of engines, and experiment condition affect the amount of exhaust gas emissions. Main disadvantage of biodiesel is increase of NOx emissions, which can be overcome by some engine modifications.

8. Conclusion

In conclusion, this literature review content covers the extensive chemical and physical properties of biodiesel and FT diesel while comparing them to mineral diesel, as well as giving an overview of their performance in CI engines. Many of the sources used in writing this report agree that biodiesel fuels are very appropriate substitute for oil fuels and are potentially suitable for reducing pollutant emissions in diesel engines, with modified combustion equipment. However, this report establishes rapeseed and soybean 30% blends to be the best alternative fuels examined here, performing within the limits demanded by the ASTM specifications and beyond the standard performance of petrol-diesel. Fischer-Tropsch also appears to be a sustainable option in the case of straight biodiesels, though POME blends between 10 and 30% performed just as well. In addition, if the current production of biomass-derived fuel was to grow, it could bring about some ethical issues and competition for produce could bring up the price of food over the globe. This could mean more starving children in third-world countries. To solve this issue, non-edible sources of oil, which can compete with edible oils, could be cultivated in non-arable lands. This would involve cultivating wasteland for biodiesel production to minimize the use of limited arable lands for growing edible oil crops for biodiesel production. Overall, further experimentation and investigation are suggested to find the most opportune solution to overcome the global oil shortage. As per ASTM standards, the calorific value of biodiesel and blends is within the range, while the flash point of pure biodiesel and its blends is slightly above that of pure diesel. Since fuel does not easily spark when exposed to flame, it is safe to handle during storage and therefore recommended for CI engines. The cetane numbers are all greater than those of pure diesel and meet ASTM standards. When burning in CI engines, Citrullus lanatus biodiesel and its blends will have the shortest possible ignition delay. A low fluidity was observed when using the fuels on CI engines, as all the pour points conformed to ASTM standards. Though fossil oils are commercially still available at low prices, and only tax credits are currently making bio-fuels a viable option, in the future this will change and research must be done to prepare for this.

Author details

Manju D. Tanwar*, Pankaj K. Tanwar, Yashas Bhand, Sarang Bhand, Kiran Jadhav and Suhas Bhand
Organic Recycling Systems (ORS) Limited, Navi Mumbai, Maharashtra, India

*Address all correspondence to: manju.tanwar@organicrecycling.co.in

References

[1] Council WE. World Energy Resources 2013 Survey, ed. WECiEa Wales. 2013

[2] Jonathan Melville GZ, Wu K. Synthesis and characterization of biofuels. UC Berkeley College of Chemistry. 2012

[3] Sadeghinezhad E, Kazi SN, Badarudin A, Oon CS, Zubir MNM, Mehrali M. A comprehensive review of bio-diesel as alternative fuel for compression ignition engines. Renewable and Sustainable Energy Reviews. 2013;**28**:410-424

[4] Arbab MI, Masjuki HH, Varman M, Kalam MA, Imtenan S, Sajjad H. Fuel properties, engine performance and emission characteristic of common biodiesels as a renewable and sustainable source of fuel. Renewable and Sustainable Energy Reviews. 2013;**22**: 133-147

[5] Subramanian KA. Biofueled Reciprocating Internal Combustion Engines. Boca Raton, FL, USA: CRC Press; 2017. p. 15

[6] IEA. Key World Energy Statistics. 2018. Available from: https://webstore.iea.org/key-world-energy-statistics-2018 [Accessed: July 12, 2020]

[7] TERM. Transport Indicators Tracking Progress towards Environmental Targets in Europe. Copenhagen, Denmark: No 7/ 2015; European Environment Agency; 2015. Available from: https://www.eea.europa.eu/publications/term-report-2015 [Accessed: July 12, 2020]

[8] IRENA. Global Energy Transformation: The REmap Transition Pathway (Background Report to 2019 Edition). Abu Dhabi, United Arab Emirates: International Renewable Energy Agency; 2019. Available from: https://www.irena.org/publications/2019/Apr/Global-energy-transformation-The-REmap-transition-pathway [Accessed: August 4, 2020]

[9] Gis W. Electromobility and hydrogeneration of the motor transport in Poland now and in the future. Journal of KONES. 2018;**25**:95-101

[10] Green Car. 2014. Available from: https://www.greencarreports.com/news/1093560_1-2-billion-vehicles-on-worlds-roadsnow-2-billion-by-2035-report [Accessed: August 4, 2020]

[11] Lamb JJ, Austbø B. Current use of bioenergy. In: Jacob J, Pollet BG, editors. Hydrogen, Biomass and Bioenergy. Trondheim, Norway, London, UK: Elsevier, Academic Press; 2020. pp. 9-20

[12] IRENA. Transport Sector Summary Charts. 2014. Available from: http://costing.irena.org/charts/transport-sector-summary-charts.aspx

[13] Torregrosa AJ, Broatch A, Plá B, Mónico LF. Impact of Fischer–Tropsch and biodiesel fuels on trade-offs between pollutant emissions and combustion noise in diesel engines. Biomass and Bioenergy. 2013;**52**:22-33

[14] Chattopadhyay S, Sen R. Fuel properties, engine performance and environmental benefits of biodiesel produced by a green process. Applied Energy. 2013;**105**:319-326

[15] Hwang J, Bae C, Gupta T. Application of waste cooking oil (WCO) biodiesel in a compression ignition engine. Fuel. 2016;**176**:20-31

[16] Xu Y, Keresztes I, Condo AM Jr, Phillips D, Pepiot P, Avedisian CT.

Droplet combustion characteristics of algae-derived renewable diesel, conventional #2 diesel, and their mixtures. Fuel. 2016;**167**:295-305

[17] Nautiyal P, Subramanian KA, Dastidar MG. Production and characterization of biodiesel from algae. Fuel Processing Technology. 2014;**120**: 79-88

[18] Viêgas CV, Hachemi I, Freitas SP, Mäki-Arvela P, Aho A, Hemming J, et al. A route to produce renewable diesel from algae: Synthesis and characterization of biodiesel via in situ transesterification of Chlorella alga and its catalytic deoxygenation to renewable diesel. Fuel. 2015;**155**:144-154

[19] Tripathi M, Sahu JN, Ganesan P, Jewaratnam J. Thermophysical characterization of oil palm shell (OPS) and OPS char synthesized by the microwave pyrolysis of OPS. Applied Thermal Engineering. 2016;**105**:605-612

[20] Krutof A, Hawboldt K. Blends of pyrolysis oil, petroleum, and other bio-based fuels: A review. Renewable and Sustainable Energy Reviews. 2016;**59**: 406-419

[21] Bezergianni S, Dimitriadis A. Comparison between different types of renewable diesel. Renewable and Sustainable Energy Reviews. 2013;**21**: 110-116

[22] Gandure J, Ketlogetswe C, Temu A. Fuel properties of biodiesel produced from selected plant kernel oils indigenous to Botswana: A comparative analysis. Renewable Energy. 2014;**68**: 414-420

[23] Bogarra-Macias M, Doustdar O, Fayad MA, Wyszynski ML, Tsolakis A, Ding P, et al. Performance of a drop-in biofuel emulsion on a single-cylinder research diesel engine. Combustion Engines PTNSS. 2016;**166**:9-16

[24] Mahmoudi H. Perfomance of Cobalt-based Eggshell Catalyst in Low Temperature Fischer Tropsch Synthesis Process to Produce Long-chain Hydrocarbons from Synthesis Gas Utilizing Fixed-bed Reactor Technology. PhD., School of Mechanical Engineering, The University of Birmingham; 2015

[25] Tristantini D, Lögdberg S, Gevert B, Borg Ø, Holmen A. The effect of synthesis gas composition on the Fischer–Tropsch synthesis over Co/γ-Al2O3 and Co–Re/γ-Al2O3 catalysts. Fuel Processing Technology. 2007;**88**(7):643-649

[26] Moazami N, Mahmoudi H, Rahbar K, Panahifar P, Tsolakis A, Wyszynski ML. Catalytic performance of cobalt–silica catalyst for Fischer–Tropsch synthesis: Effects of reaction rates on efficiency of liquid synthesis. Chemical Engineering Science. 2015;**134**:374-384

[27] Maniatis K, Millich E. Energy from biomass and waste: The contribution of utility scale biomass gasification plants. Biomass and Bioenergy. 1998;**15**(3): 195-200

[28] Zeng S, Du Y, Su H, Zhang Y. Promotion effect of single or mixed rare earths on cobalt-based catalysts for Fischer–Tropsch synthesis. Catalysis Communications. 2011;**13**(1):6-9

[29] Hessam Jahangiri J B, Parvin Mahjoubi, Karen Wilson, Sai Gu, A review of advanced catalyst development for Fischer–Tropsch synthesis of hydrocarbons from biomass derived syn-gas. The Royal Society of Chemistry, 2014

[30] Rafiq MH, Jakobsen HA, Schmid R, Hustad JE. Experimental studies and modeling of a fixed bed reactor for

Fischer–Tropsch synthesis using biosyngas. Fuel Processing Technology. 2011;**92**(5):893-907

[31] Moazami N, Mahmoudi H, Panahifar P, Rahbar K, Tsolakis A, Wyszynski ML. Mathematical modeling and performance study of fischer-tropsch synthesis of liquid fuel over cobalt-silica. Energy Procedia. 2015;**75**:62-71

[32] Henrici-Olivé G, Olive S. Mechanism of the fischer-tropsch synthesis: Origin of oxygenates. Journal of Molecular Catalysis. 1984;**24**(1):7-13

[33] Moazami N, Wyszynski ML, Mahmoudi H, Tsolakis A, Zou Z, Panahifar P, et al. Modelling of a fixed bed reactor for Fischer–Tropsch synthesis of simulated N2-rich syngas over Co/SiO2: Hydrocarbon production. Fuel. 2015;**154**:140-151

[34] Robert Guettel UK. Thomas Turek, Reactors for Fischer-Tropsch Synthesis. Institute of Chemical Process Engineering: Clausthal University of Technology. Chem. Eng. Technol; 2008

[35] Liu G et al. Making Fischer—Tropsch fuels and electricity from coal and biomass performance and cost analysis. Energy & Fuels. 2010;**25**:415-437

[36] Basha SA, Gopal KR, Jebaraj S. A review on biodiesel production, combustion, emissions and performance. Renewable and Sustainable Energy Reviews. 2009;**13**(6–7):1628-1634

[37] Hong IK, Jeon GS, Lee SB. Prediction of biodiesel fuel properties from fatty acid alkyl ester. Journal of Industrial and Engineering Chemistry. 2014;**20**(4): 2348-2353

[38] Knotthe G, Krahl J, Gerpen J. The Biodiesel Handbook. 2nd ed. 1 May 2010. eBook ISBN: 9780983507260

[39] Knothe G. Biodiesel and renewable diesel: A comparison. Progress in Energy and Combustion Science. 2010;**36**(3): 364-373

[40] Biomass-Biodiesel Handling and Use Guidelines. 3rd ed. U.S. Department of Energy, DOE/GO-102006-2358. Sep 2006

[41] Elgharbawy AS, Sadik WA, Sadek OM, Kasaby MA. A review on biodiesel feedstocks and production technologies

[42] Maru MM, Trommer RM, Cavalcanti KF, Figueiredo ES, Silva RF, Achete CA. The Stribeck curve as a suitable characterization method of the lubricity of biodiesel and diesel blends. Energy. 2014;**69**:673-681

[43] Hazrat MA, Rasul MG, Khan MMK. Lubricity improvement of the ultra-low sulfur diesel fuel with the biodiesel. Energy Procedia. 2015;**75**:111-117

[44] Jääskeläinen H. DieselNet Technology Guide: Fuel Property Testing: Lubricity, 2008. Available from: https://www.dieselnet.com/tech/fuel_diesel_lubricity.php

[45] Suarez PAZ, Moser BR, Sharma BK, Erhan SZ. Comparing the lubricity of biofuels obtained from pyrolysis and alcoholysis of soybean oil and their blends with petroleum diesel. Fuel. 2009;**88**(6):1143-1147

[46] Energy CR. Biodiesel fuel specifications and comparison to diesel fuel. CrimsonRenewable.com. 2015

[47] Lapuerta M, Sánchez-Valdepeñas J, Sukjit E. Effect of ambient humidity and hygroscopy on the lubricity of diesel fuels. Wear. 2014;**309**(1–2):200-207

[48] McCarthy P, Rasul MG, Moazzem S. Analysis and comparison of performance

and emissions of an internal combustion engine fuelled with petroleum diesel and different bio-diesels. Fuel. 2011;**90**(6): 2147-2157

[49] Sivasamy A, Cheah KY, Fornasiero P, Kemausuor F, Zinoviev S, Miertus S. Catalytic applications in the production of biodiesel from vegetable oils. ChemSusChem. 2009;**2**(4):278-300

[50] Steidley GK a KR. Lubricity of components of biodiesel and petrodiesel. The Origin of Biodiesel Lubricity, Energy and Fuel American Chemical Society. 2005;**19**:8

[51] Missouri. Energy for Missouri: Today and Tomorrow - Educator's Guide. Energy Producing Systems Biomass Power. Available from: http://dnr.mo.gov/education/energy/docs/biomass power.pdf

[52] Sukjit E. Synergistic Effects of Alcohol Based Renewable Fuels: Fuel Properties and Emissions. PhD., School of Mechanical Engineering, The University of Birmingham; 2013

[53] Sukjit E, Dearn KD, Tsolakis A. Interrogating the surface: The effect of blended diesel fuels on lubricity. SAE International Journal of Fuels and Lubricants. 2011;**5**(1):154-162

[54] Magin Lapuerta RGa-C, Agudelo JR. Lubricity of ethanol-biodiesel-diesel fuel blends. Energy & Fuels. 2009;**24**:6

[55] Datta A, Mandal BK. A comprehensive review of biodiesel as an alternative fuel for compression ignition engine. Renewable and Sustainable Energy Reviews. 2016;**57**:799-821

[56] Gumus M, Kasifoglu S. Performance and emission evaluation of a compression ignition engine using a biodiesel (apricot seed kernel oil methyl ester) and its blends with diesel fuel. Biomass and Bioenergy. 2010;**34**(1): 134-139

[57] Gill SS, Tsolakis A, Dearn KD, Rodríguez-Fernández J. Combustion characteristics and emissions of Fischer–Tropsch diesel fuels in IC engines. Progress in Energy and Combustion Science. 2011;**37**(4):503-523

[58] Rao GLN, Prasad BD, Sampath S, Rajagopal K. Combustion analysis of diesel engine fueled with Jatropha oil methy Lester-diesel blends. International Journal of Green Energy. 2007;**4**(6): 645-658

[59] Laforgia D, Ardito V. Biodiesel fueled IDI engines: Performances, emissions and heat release investigation. Bioresource Technology. 1995;**51**(1): 53-59

[60] Canakci M, Van Gerpen JH. The performance and emissions of a diesel engine fuelled with biodiesel from yellow grease and soybean oil. In: 2001 ASAE Annual International Meeting. The Society for Engineering in Agrcultural, Food and Biological Systems; 2001. Available from: http://web.cals.uidaho.edu/biodiesel/files/2013/08/ASABE-016050.pdf

Section 2

Biodiesel Transesterification

Chapter 2

Direct Transesterification: From Seeds to Biodiesel in One-Step Using Homogeneous and Heterogeneous Catalyst

Issis Claudette Romero-Ibarra, Araceli Martínez Ponce Escuela, Gabriela Elizabeth Mijangos Zúñiga and Wendy Eridani Medina Muñoz

Abstract

Biodiesel is a renewable alternative biofuel and is an option to diversify the conventional fossil fuels. Moreover, biodiesel is nontoxic, biodegradable, and biomass renewable diesel fuel and its combustion produces low amount of CO, CO_2, hydrocarbon, and particulate matter. It can be produced through transesterification reaction. The most common method is homogeneous transesterification process using basic catalyst as NaOH. However, this route has drawbacks as long timespans, saponification reaction, a large amount of solvent, and a large amount of water to neutralize the methyl esters to eliminate the catalyst. This chapter presents the direct transesterification as a green and sustainable alternative method to improve the benefits of conventional transesterification. The direct transesterification is a one-step process to obtain biodiesel from seed crops in presence of a catalyst. *Jatropha curcas* L. and *Ricinus communis* have been evaluated as non-edible seeds feedstocks. Also, various acid and basic homogeneous and heterogeneous catalysts have been investigated. Results shown that heterogeneous direct transesterification yields ~99% with 5 wt% catalyst in 4 h without n-hexane for oil extraction or water for purify the biodiesel. Heterogeneous direct transesterification is a promising method of obtaining biodiesel as methanol acts as a reactant and as a solvent.

Keywords: direct transesterification, homogeneous catalyst, heterogeneous catalyst, biodiesel, *Jatropha curcas* seeds, *Ricinus comunis* seeds

1. Introduction

The production of biofuels from renewable biomass resources is an attractive way to mitigate CO_2 emissions and alleviate the shortage of fossil fuels. Biofuels have emerged as one of the most promising renewable energy sources, offering one alternative to substitute for petroleum-based diesel fuel. Biomass offers an alternative solution, low-cost solution to a drop-in transportation fuel for blending with conventional diesel. The most common biofuels are based on vegetable oil for fatty acid methyl esters (FAMEs) biodiesel. However, the use of edible oils for bio-based fuels production may create controversy. The biodiesel production from edible oil crops has been questioned in several fields, and their negative effects by its high cost and their destruction of soil cultivation are discussed. Recently, agro-industrial waste and sustainably managed non-food feedstocks have been considered as a new generation of fatty acid methyl esters (FAMEs) or advanced biodiesel. For example, non-edible oilseed crops such as *Ricinus communis* oil and *J. curcas L.*. *R. communis* are a new option for biodiesel production because they are not edible, due to their high oil content, easy propagation, resistance, rapid growth, and adoption in wide agroclimatic conditions [1–7], and it provides commercially viable alternative to edible oils. These oils are not suitable for human consumption [8], and therefore, its production does not interfere with the food industry or harvesting lands.

Biodiesel from renewable biomass resources is an attractive way to mitigate CO_2 emissions and alleviate the shortage of fossil fuels [4–6, 9, 10]. Biodiesel can be produced *via* esterification or transesterification process. Triglycerides (TG) from vegetable oils and animal fats react with short-chain alcohols and acid or base catalysts to render fatty acid methyl esters (FAMEs) and glycerol as by-product [9, 10]. Every triglyceride molecule reacts with three equivalents of alcohol to produce glycerol and three fatty acid (methyl) ester molecules. Methanol is the most common alcohol due to its low price, high activity, and green chemistry metrics. Nowadays, the biodiesel is produced industrially by homogeneous transesterification due to the high yields, lower cost, and the short reaction time. The most common homogeneous catalysts used are NaOH and KOH as basic catalyst and H_2SO_4 as an acid catalyst. However, homogeneous acid and base catalysts present many disadvantages such as long timespans, a large amount of solvent, reactors and engine manifolds corrosion, the difficulty of phase separation, their removal from the resulting biodiesel from the reaction, the neutralization and purification of the biofuel at the end of the reaction, and the excessive use of water to removal emulsions and soaps for these processes. The base catalysts cause saponification, a competition reaction, and decrease the yield of the biodiesel obtained. The acid catalysts increase the biodiesel yield, but stable conditions are needed, or the reaction is affected and slows down the transesterification. As by-product, glycerol has significant value to the industry. However, the recovered glycerol is normally impure due to the presence of salts, soaps, monoglycerides, and diglycerides, and the glycerol purification process consequently adds an additional cost [11, 12].

On the other hand, the use of heterogeneous catalysts represents another strategy to obtain biofuels. These catalysts seek to make the production of biodiesel greener, generating less aggressive residues to the environment and reducing manufacturing costs. In transesterification reaction with heterogeneous catalyst, the reagents and the catalyst are in different phases. Usually, the catalyst is solid; meanwhile, the reagents are in liquid phase. The use of heterogeneous catalysts would result in simpler, cheaper separation processes, a reduced water effluent load as well as lower energy consumption and cleaner operating conditions [9].

Furthermore, catalyst would not have to be continuously added and would be easier to reuse. In general, the saponification reaction is not common with heterogeneous catalyst, and the purity of the products increases because the catalyst is more selective. There would be no neutralization products, enhancing the purity of the glycerol by-product [11, 13]. Several solid base and acid catalysts are explored in transesterification reaction because they offer enhanced process, avoid the quenching steps, and allow continuous operation. Solid acid as heterogeneous catalyst in transesterification reaction of oils into biodiesel renders lower activity, also it is necessary for higher reaction temperatures, and they are lower in comparison with base catalyst [14]. In contrast to solid bases catalysts, solid acids can esterify free fatty acids (FFAs) though to FAME. Some examples of acid heterogeneous catalysts are ZrO_2, TiO_2, SnO_2, zeolites, etc., whereas the basic heterogeneous catalysts most commonly used are CaO, MgO, SrO. The yields obtained with these basic catalysts were 97%, 92%, and 98%, respectively [2, 3, 15, 16]. Some of these catalysts are used to obtain biodiesel by means of used cooking oil, algae, and different oils such as palm oil [13, 15–19]. CaO/TiO_2 using canola oil results in a yield of 96.9% [18]. Another alternative is the use of biomass, for example, papaya seed ash as a catalyst, as it contains metal oxides such as K_2O, MgO, and CaO having a yield of 95.6% [19]. Besides, it was shown that sodium zirconate (Na_2ZrO_3) exhibited interesting catalytic properties as a basic heterogeneous catalyst to produce biodiesel *via* a soybean oil transesterification reaction, which led to good purity [20–22]. Furthermore, the alteration of the chemical composition as a possible increment of active basic sites, incorporating cesium to the Na_2ZrO_3 catalyst by a simple impregnation method, improves the yield of reaction due to its basic character. The catalytic system Cs-sodium zirconate (cesium-impregnated sodium zirconate) was considered as modified in transesterification reaction of soybean oil and jatropha oil for biodiesel production [23]. Results showed outstanding FAME conversion, from soybean oil, of 98.8% using 1 wt% of catalyst in just 15 min. These catalysts can operate at low concentrations and in short reaction times due to its high basicity and low solubility [20, 23]. Recently, other ceramic materials such as sodium silicate (Na_2SiO_3) [24] and sodium zincsilicate (Na_2ZnSiO_4) [25] have been evaluated with high yields to produce biodiesel $\sim$99% and $\sim$ 98%, respectively.

However, these conventional processes consist of two steps: first, the oil extraction and then transesterification of triglycerides in the presence of the catalyst. Both steps generate drawbacks that promote a more expensive biodiesel production. Therefore, for both environmental and economic reasons, there is increasing interest in new alternatives to produce biofuels. It will face new challenges as a technology to overcome the feasible routes to biofuels. Hence, green chemistry presents a new synthesis route to obtain products from conventional or traditional methods favoring the sustainable process, avoiding hazardous and toxic substances (e.g., hexane or dichloromethane) to extract vegetable oils, and favoring the energy efficiency. As a new alternative, the direct transesterification represents an innovative and *in situ* route to obtain biodiesel from non-edible oils-seed in one step in the presence of homogeneous and heterogeneous catalysts [21, 22].

2. Direct transesterification of seeds

It is well known that biodiesel is conventionally produced by transesterification of vegetable oils using an alcohol and homogeneous or heterogeneous catalysts. The conventional transesterification of oils is based on two steps: 1) the oil extraction and then 2) the transesterification of triglycerides (**Figure 1**) [15, 16, 26, 27]. In the first step, there are several basic methods for obtaining oils, such as chemical extraction, supercritical fluid extraction, steam distillation, mechanical extraction. The second

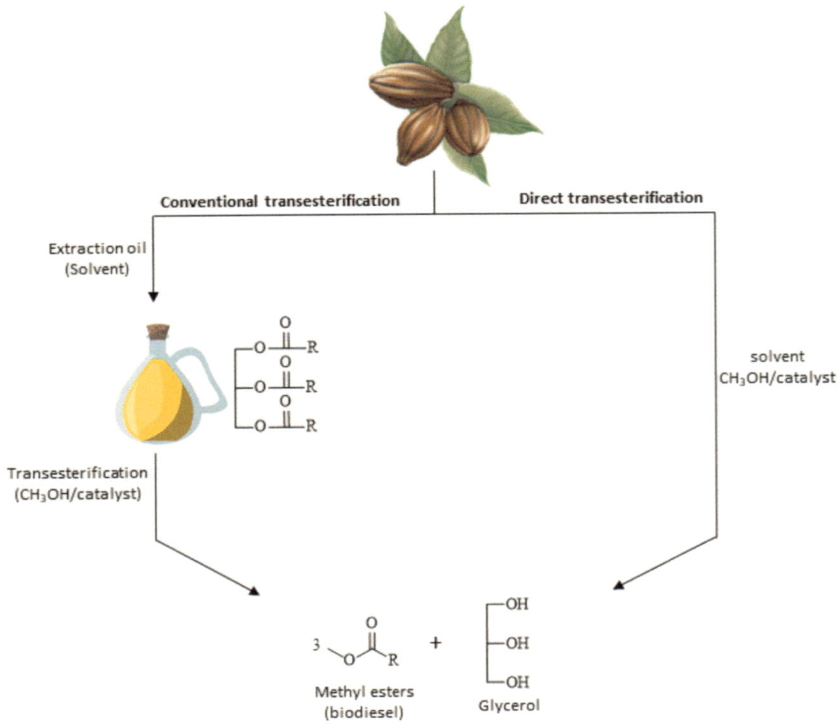

Figure 1.
Conventional and direct transesterification of seed oil.

step is the transesterification of triglycerides from homogeneous or heterogeneous catalysts. Both catalysts could be acid or basic. For example, biodiesel production through the conventional transesterification from soybean, sunflower, and *Cynara cardunculus* oils (edible oils) and Jatropha and *R. communis* oil (non-edible oils), with methanol or ethanol using acidic (HCl, H_2SO_4) and basic (NaOH, KOH) homogeneous catalysts, has been reported [5, 26–28].

An alternative to the conventional method is the direct transesterification (**Figure 1**). **Figure 1** shows the comparison between conventional (two steps) and direct transesterification (one step). Direct transesterification is a novel green one-pot synthesis to produce biodiesel from non-edible seeds using homogeneous or heterogeneous catalysis. In this *in situ* one-step process, the extraction and transesterification take place simultaneously in one vessel. The one-step process is more economical and efficient than conventional method [21, 22]. Recently, a few studies have demonstrated that biodiesel production can be achieved from the direct transesterification of seed oils or microalgae biomass with homogeneous catalysts [28–34]. Although both alkaline and acid homogeneous catalysts were suitable for the reaction, most studies use a basic catalyst (sodium hydroxide, potassium hydroxide, or sodium methoxide) due to reduced corrosiveness, lower amount of catalyst, and reaction time. Also, biodiesel production through the homogeneous direct transesterification from edible oils such as soybean, sunflower, and *Cynara cardunculus* seeds, and non-edible oils such as *Jatropha curcas L* and *Ricinus communis* seeds, with methanol or ethanol, has been reported [1, 5, 28, 31]. FAME yields are very high and comparable with conventional homogeneous as well as heterogeneous catalysis (~99%).

Recently, heterogeneous direct transesterification represents a new alternative in biodiesel production; due to these, solid catalysts have many advantages over liquid catalyst used in homogeneous direct transesterification. The heterogeneous catalyst exhibits easy separation, lower energy consumption, and cleaner operation. Using solid catalyst could be eliminated the contaminated waste, the formation of soaps, and the emulsification of products that are generated in the homogeneous acid and basic transesterification [21, 22]. It is important to note that the heterogeneous reaction can proceed without polluting or hazardous solvents, for example, without *n*-hexane for oil extraction. Thus, methanol as a reactant acts as a solvent to carry out the reaction and obtain the methyl esters. This important solvent change increased in seven times the greenness of the heterogeneous reaction in comparison with the conventional method (two-step). Besides, the water consumption to purify the biodiesel is not necessary. Direct transesterification shows the environmental benefits related to solvents and energy consumption [21, 22].

2.1 Direct transesterification of *J. curcas* and *R. communis* using homogeneous catalyst

2.1.1 Seed-oil extraction

J. curcas L. is a plant oil with more than 3500 species. It is native to Mesoamerica, covering northern Mexico and Central America. Jatropha seed contains a high percentage of vegetable oil that can be used in biodiesel production [28, 30, 35]. Another vegetable oil used for biodiesel production is the *R. communis*, which is native to tropical Africa and is considered a highly invasive species in Some Asian and European countries [5, 21, 22]. *J. curcas L.* and *R. communis* seeds oils are suitable for biodiesel production because they are highly available seeds and neither their fruits nor plants are edible. These seeds are widely distributed in several places as a weed in urban and agricultural areas, and therefore, they have a great capacity for adaptation that allows them to be cultivated in all tropical and subtropical regions, although it is typical in semiarid regions [32–35]. Both species are considered toxic plants in the human. For example, jatropha seed contain toxic compounds known as phorbol esters, while in the *R. communis* seed, there is an albumin known as ricin. These toxic compounds can cause diarrhea, rapid breathing, tumor promotion in humans, etc., in high concentrations [36]. The extraction of the oil is carried out by several methods such as chemical extraction, supercritical fluid extraction, steam distillation, mechanical extraction, solvent extraction, CO_2 extraction, maceration, enfleurage, among others [30, 37, 38]. Also, seed oils can be obtained through the Soxhlet extraction method [37]. These methods of extraction require an excessive energy consumption so in the most cases the cost of the final product becomes more expensive. In addition, the extraction of the vegetable oil involved the use of toxic solvents and water, which makes the process less environmentally friendly.

Table 1 shows the results of jatropha and *R. communis* oils extraction from the seed and the shell, using *n*-hexane and methanol as solvents. These seeds were provided by the State of Morelos in Mexico. The oil content of jatropha and *R. communis* seeds corresponded to yields ranging from 48 to 52% and from 50 to 52%, respectively (*entries 1–4*), while the oil content of the shell was only <7% (*entries 5 and 6*). It is important to know that the quantity of the seed oils helps to quantify the amount of biodiesel in the direct transesterification of seeds.

Entry[a]	*Jatropha curcas L.*	*R. communis*	Solvent	Yield %
1	Seed	—	*n-hexane*	52
2	Seed	—	Methanol	48
3	—	Seed	*n-hexane*	52
4	—	Seed	Methanol	50
5	Shell	—	*n-hexane*	5
6		Shell	*n-hexane*	7

[a]*Soxhlet extraction method for 6 hours.*

Table 1.
Ricinus communis *and jatropha oils extraction.*

The amount of vegetables oils extracted agreed with the values reported in the literature, which ranges from 40 to 60% for jatropha oil [1, 28, 36–39] and ranges from 40 to 56% for *R. communis* oil [39] by the chemical oil extraction method (Soxhlet extraction). Therefore, the percentage of the vegetable oil extraction in both seeds with n-hexane and methanol is similar. Therefore, the methanol is an interesting option to carry out the reactions as a reactant and solvent.

Vegetable oils (triglycerides) contain mainly mixture of triglycerides (TAG), with a different composition of the alkyl chains depending on their origin. **Figure 2** shows the jatropha (**Figure 2A**) and *R. communis* (**Figure 2B**) oil compositions that were calculated according to electrospray ionization mass spectrometry (ESI-MS) analysis [40]. Jatropha

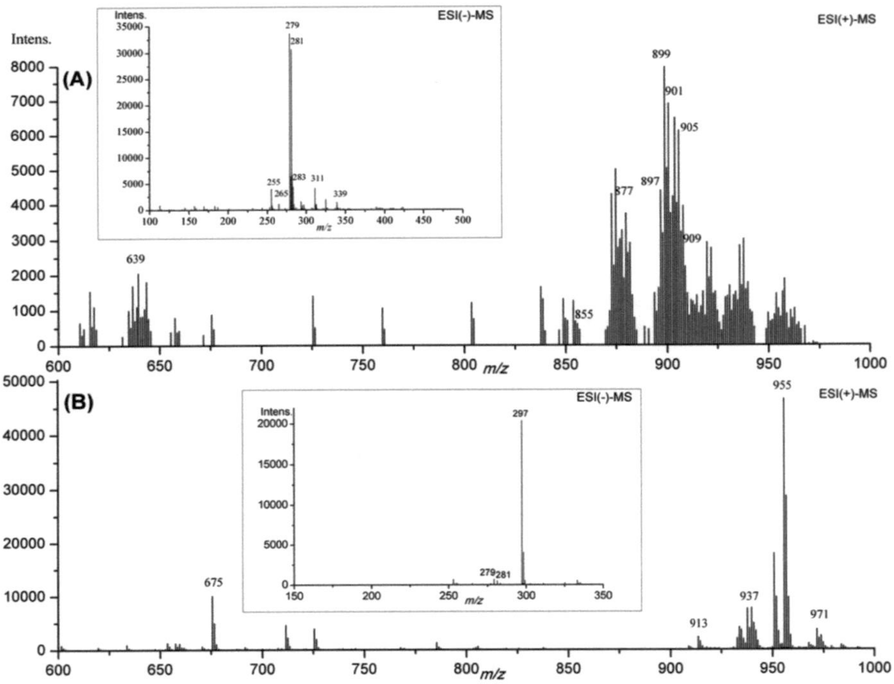

Figure 2.
TAG and FFA fingerprints of jatropha (A) and Ricinus communis *(B) oils obtained by ESI(+)-MS and ESI (−)-MS ionization technique.*

oil mainly contains oleic (O), linoleic (L), and linolenic (L_n) triglycerides. The most abundant $[TAG + Na]^+$ ions are of *m/z* 905 ($C_{57}H_{102}O_6$), 901 ($C_{57}H_{98}O_6$), and 899 ($C_{57}H_{96}O_6$) that correspond to LOO, LLL, and L_nLL, respectively (**Figure 2A**). Other triglycerides of *m/z* 907 ($C_{57}H_{104}O_6$), 903 ($C_{57}H_{100}O_6$), and 897 ($C_{57}H_{94}O_6$) are attributed to OOO, OLL, and L_nL_nL in less amount. Diglyceride as LL (*m/z* 639, $C_{39}H_{68}O_5$) was found. The free fatty acids (FFAs) contained in Jatropha oil were detected by ESI(−)-MS. **Figure 2A** shows the spectrum of ESI(−)-MS that displays mainly ions corresponding to the deprotonated molecules $[FFA\text{-}H]^-$ from oleic (*m/z* = 281, $C_{18}H_{34}O_2$), linoleic (*m/z* = 279, $C_{18}H_{32}O_2$) acids as the most abundant. In the same way, *R. communis* oil contains several lipids such as tri-, di-glycerides, and FFA. The $[TAG+Na]^+$ ion most abundant is the ricinoleic TAG (RRR) with *m/z* = 955 ion. Diglycerides such as RR (m/z, C39:2) and ricinoleic acid (*m/z* = 297, C18:0) were detected (**Figure 2B**).

2.1.2 Conventional and direct transesterification processes

The production of biodiesel from vegetable oils and fats can be carried out by several routes (pyrolysis, microemulsion, transesterification, etc.). The most used conversion method is from the transesterification of triglycerides in the presence of homogeneous basic catalysts (NaOH and KOH) with methanol. It has been reported that by this synthesis route, the conversion of oil to biodiesel is 99.99%. One of the disadvantages at the industrial level is the recovery of the catalyst and the emulsions and soaps that are obtained during the reaction. Moreover, this conventional process consists of two steps: First, it is necessary extract the vegetable oil from different seeds prior to the transesterification, commonly the oil is extracted with a hazardous solvent or mechanical extraction with high energy consumption and then, in a second step, subsequently transesterification of triglycerides in the presence of the catalyst. These disadvantages promote excessive energy consumption and high operational and biodiesel production costs. Therefore, a new strategy for obtaining biodiesel is proposed. Direct transesterification is a method (*in situ* or one-pot) for transforming seed oils (biomass) to free acid alkyl esters and glycerin, in the presence of a short-chain alcohol and the catalyst. This one-step process is more economical and efficient than conventional method. **Figure 1** shows the comparison between conventional (two steps) and direct transesterification (one step).

Table 2 shows the conventional and direct transesterification reactions of jatropha and *R. communis* oils using acidic (HCl) (*entries 1 and 9*) and basic (NaOH) homogeneous catalysts.

The Jatropha and *R. communis* oils were transesterified in methanol (CH_3OH) using acidic catalyst (HCl/CH_3OH, 5 *v/v* %) that corresponds to (HCl/oil, 31 *wt.* %); the oil conversion to fatty acid methyl esters (FAME's) was 98.70% and 99.10%, respectively (*entries 1 and 9*). In the same way, these non-edible oils were transesterified using basic catalyst ($NaOH/CH_3OH$, 5 *wt.* %) that corresponds to (NaOH/oil, 1.20 *wt.* %) with 99.99% of conversion to FAME.

It has been reported that the use of acidic homogeneous catalysts can catalyze esterification and transesterification simultaneously; however, it is not sensitive to the free fatty acids (FFA) content in the oils, the oil conversion to FAME needs high catalyst concentration, long reaction times, high molar ratio of alcohol to oil, and the catalyst separation is difficult in comparison with the base-catalyzed process [41–47]. In **Table 2**, the vegetable oil transesterification using the acidic catalyst (HCl) was conducted by conventional reflux for 6 hours, while for the basic catalyst (NaOH), the reaction remained at 50°C for 3 hours (*entry 10*). Therefore, the direct

Entry	Seed	Catalyst	Catalyst/oil (wt. %)	[CH_3OH]/[Oil][b]	Time (h)	Solvent	FAME Yield[c] (%)
1[a]	Jatropha	HCl	31.00	327:1	6	—	98.70
2[a]	Jatropha	NaOH	1.20	9:1	3	—	99.99
3	Jatropha	NaOH	0.30	9:1	9	*n-hexane*	45.24
4	Jatropha	NaOH	0.30	16:1	9	*n-hexane*	54.53
5	Jatropha	NaOH	1.20	9:1	9	*n-hexane*	81.16
6	Jatropha	NaOH	1.20	16:1	9	*n-hexane*	99.99
7	Jatropha	NaOH	1.20	65:1	9	methanol	99.99
8	Jatropha	NaOH	2.00	16:1	9	*n-hexane*	99.99
9[a]	*R. communis*	HCl	31.00	327:1	6	—	99.10
10[a]	*R. communis*	NaOH	1.20	9:1	3	—	99.99
11	*R. communis*	NaOH	0.30	9:1	9	*n-hexane*	48.29
12	*R. communis*	NaOH	0.30	16:1	9	*n-hexane*	53.00
13	*R. communis*	NaOH	1.20	16:1	9	*n-hexane*	99.99
14	*R. communis*	NaOH	1.20	37:1	9	methanol	97.80
15	*R. communis*	NaOH	2.00	16:1	9	*n-hexane*	99.99

[a]*Conventional transesterification.*
[b]*Molar ratio of methanol to oil.*
[c]*Isolated fatty methyl esters, FAME.*

Table 2.
Biodiesel from conventional and direct transesterification of jatropha and Ricinus communis *seed oils using homogeneous catalysts.*

transesterification of seed oils was carried out from the basic homogeneous catalyst, NaOH. The solvent extraction step that is required in the conventional process but not in direct transesterification, and is usually the most capital and running cost-intensive [46]. Few studies have demonstrated the FAME production achieved from the *in situ* transesterification using a homogeneous catalyst, also reactive extraction, of several seed oils such as soybean [30], sunflower [1, 31], Jatropha [22, 28], *R. communis* [21], microalgae (*Schizochytrium limacinum, Chlamydomonas, and Chlorella*) [32, 33], and others. In this method, oil-bearing seeds are ground and then reacted directly with the alcohol and catalyst (basic homogeneous and heterogeneous catalysts), thereby eliminating the timespans and a large amount of solvent for the oil extraction.

In particular, the homogeneous direct transesterification of the jatropha and *R. communis* seed oils in hexane or methanol as solvents, in the presence of NaOH basic catalyst, was studied. **Table 2** shows the results of the jatropha (*entries 3–8*) and *R. communis* (*entries 11–15*) seed oil conversion to free acid methyl esters (FAMEs, biodiesel) to different conditions of reaction (ratio of catalyst to oil, molar ratio of CH_3OH to oil, and time). The effect of several factors that are type of solvent, catalyst concentration, temperature, reaction time, methanol-oil ratio, and particle size has been investigated to optimize the direct transesterification of seed oils for achieving maximum oil yield [33, 48–50]. Jatropha and *R. communis* seed oil conversion to FAME from direct transesterification was optimized using the following parameters: NaOH homogeneous catalyst amount, the ratio of methanol/oil, and the effect of reaction, which are described below.

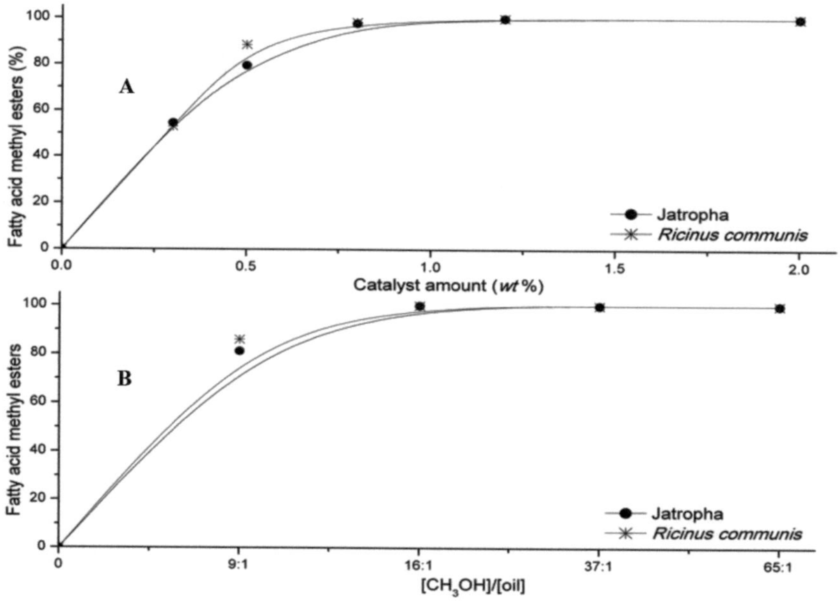

Figure 3.
FAME yield for the direct transesterification of jatropha and Ricinus communis *seed oil* versus *catalyst amount (**A**) and ratios [CH_3OH]/[oil] (B).*

Figure 3 shows the FAME yield obtained from homogeneous direct transesterification of jatropha and *Ricinus communis* seeds. **Figure 3A** shows the FAME yield *versus* catalyst amount, and **Figure 3B** shows the relation between FAME yield *versus* molar ratio [CH_3OH]/[oil].

1. Effect of NaOH amount

   Figure 3A shows the results of direct transesterification reactions of jatropha and *Ricinus communis* seeds using methanol and NaOH as catalyst with ratio of NaOH/oil = 0.3 *wt.* % to 1.2 *wt.* %. The molar ratio [CH_3OH]/[oil] = 16:1, time = 9 hours were maintained constant using hexane as solvent (**Table 2**, *entries 4 and 12*). The catalyst amount affects the yield of the FAME products for both seeds. Low catalyst concentration (NaOH/oil = 0.3% by weight) reached the maximum yield for jatropha and *R. communis* seed oil of 54.53% and 53.00%, respectively. The conversion increased as the amount of catalyst increased from 0.5 *wt.* % to 0.8 *wt.* %, varying for the jatropha seed from 79.5% to ~97.7% and for the *R. communis* seed from 88.45 to 98.2%, respectively. Then, the maximum yield reached an equilibrium. The optimum catalyst amount for both seed oils was at 1.2 *wt.* % of catalyst, and the FAME yield was ~99.99%. An increment in catalyst amount (2 *wt.* %) does not affect the oil conversion to FAME.

2. Effect of methanol-oil ratio

   Figure 3B shows the plot of methanol-oil ratio *versus* FAME yield of the direct transesterification of jatropha and *Ricinus communis* using a molar ratio

NaOH/oil = 1.2 *wt.* % (**Table 2**, *entries 5–7, 13,* and *14*). The methanol-to-oil molar ratio varied for jatropha and *R. communis* seed oils within the range of 9:1–65:1. The maximum oil conversion to FAME products (99.99%) in both seed oils was obtained at the methanol-to-oil molar ratio of 16:1. **Table 2** shows the results of these reactions when *n*-hexane is used as solvent (*entries 5, 6,* and *13*). The excess methanol in the direct transesterification (methanol-to-oil molar ratio of 37:1 and 65:1) is used as a solvent, and the conversion of FAME was 99.99% (**Table 2**, *entries 7* and *14*).

3. Influence of the reaction time

Figure 4 shows the kinetic curves of FAME yields *versus* reaction time of the jatropha and *Ricinus communis* seeds. According to the previous results, the optimum catalyst was 1.2 *wt.* % of NaOH and a molar ratio [CH_3OH]/[oil] = 16:1. The oil conversion to FAME jatropha and FAME *R. communis* was increased 65% and 64% from 0.5 hours, respectively. Over the period from 1 to 8 hours, FAME yield from jatropha seed was increased with values ranging from 71 to 98% and for *R. communis* seed was from 68.5% to 98.4%, respectively. The reactions reached equilibrium after 9 hours with the maximum conversion of 99.99% in both cases. The maximum conversion to FAME jatropha and FAME *R. communis* is observed in the ^{1}H-NMR spectra (**Figure 5**).

Figure 5 shows the ^{1}H-NMR spectra of FAME products obtained from jatropha seeds (A) and FAME products obtained from *Ricinus communis* (B). The signal arising in 3.66 ppm region corresponds to the protons of CH_3-O-. The signals observed at 5.39–5.30 ppm (A) and 5.59–5.53 ppm to 5.42–5.34 pp. (B) represent the protons of

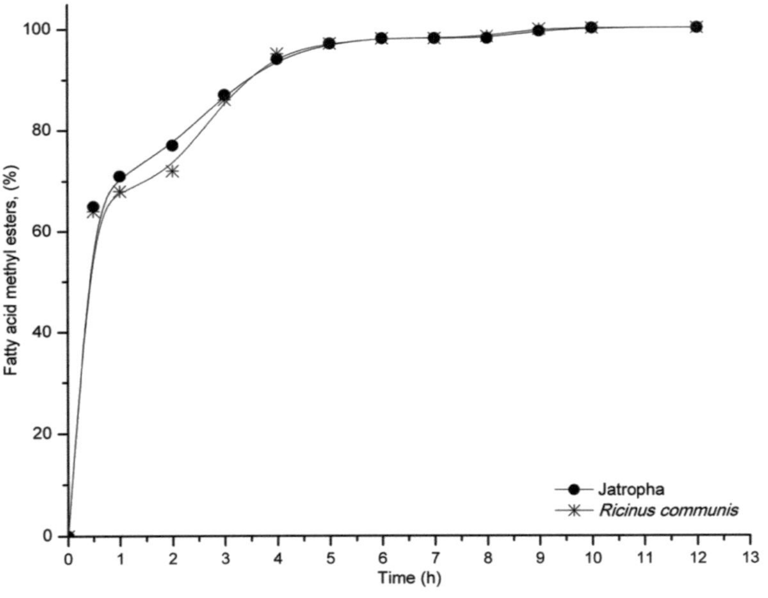

Figure 4.
Reaction time influence on FAME yields (direct transesterification) from Jatropha and Ricinus communis *seeds.*

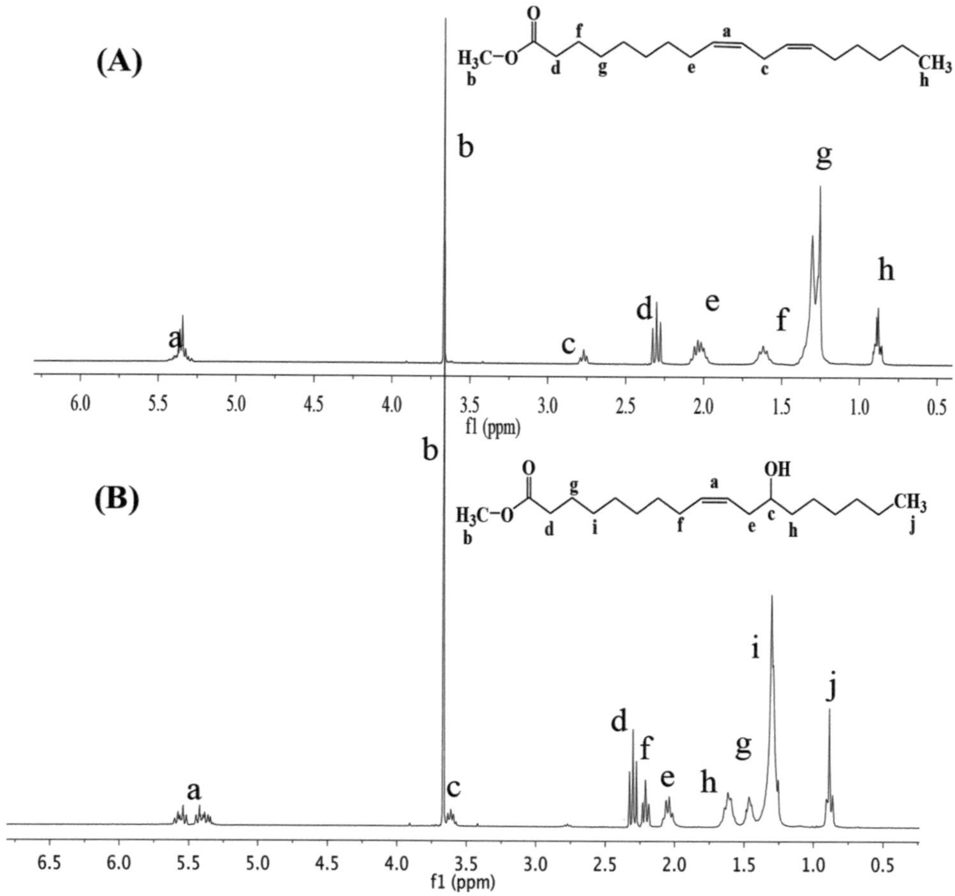

Figure 5.
^{1}H-NMR (400 MHz, $CDCl_3$) spectra of jatropha FAME (A) and Ricinus communis *FAME (B).*

vinyl group (HC=CH) contained in the unsaturated fatty acids. The FAME obtained from jatropha; the spectrum (A) shows an intense signal at 2.79 ppm compared with the spectrum (B) that corresponds to protons of -CH_2 group bonded to carbon-carbon double bond in the methyl linoleate chain. The FAME *R. communis* is composed of methyl ricinolate chains. The spectrum (B) shows the signal of the hydroxyl group (-CH-OH) that is observed at 3.65–3.59 ppm.

The composition of the Jatropha and *R. communis* biodiesel products obtained from direct transesterification of seed oil using NaOH/MeOH catalysts was determined from electrospray ionization mass spectrometry (ESI-MS) technique. The biodiesel composition was detected as (FAME + Na]$^+$ ions. **Figure 6A** shows the chromatogram of fatty acid methyl esters (FAMEs) from the seed of *Jatropha curcas L.* We can see that the oleic- (C18:1, 30.00%) and linolenic (C18:2, 48.30%) methyl esters compounds are the most abundant, respectively. The palmitic- (C16:1), stearic- (C18:0), and eicosanoic methyl esters (m/z = 349, C20:0) signals are shown in low ratio. **Figure 6B** shows the chromatogram of FAMEs from the seed of *R. communis*. The most abundant signal corresponds to the ricinoleic methyl ester (C18:1-OH, 91.50%),

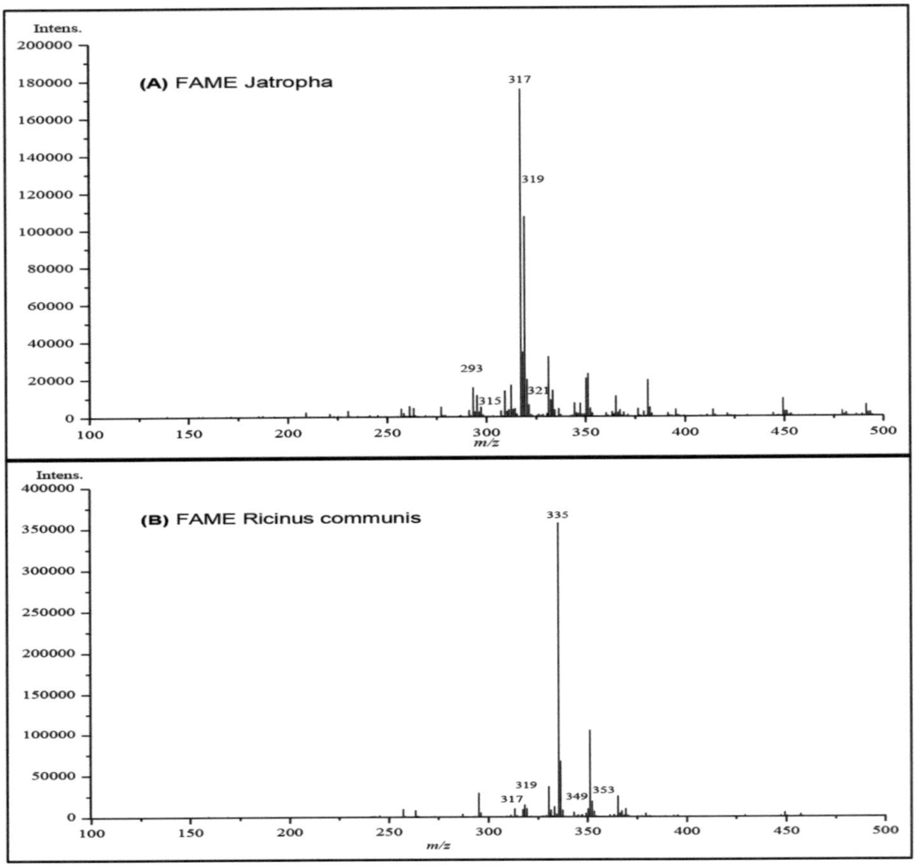

Figure 6.
ESI (+)-MS of fatty acid methyl esters (FAME) from (A) Jatropha curcas *and (B)* Ricinus communis *seeds.*

followed by FAMEs in smaller amounts such as linoleic- (C18:2), oleic- (C18:1), stearic- (C18:0), eicosanoic- (C20:0), and dihydroxystearic (C18:0-(OH)2) methyl esters, respectively.

2.2 Direct transesterification of *J. curcas* and *R. communis* using heterogeneous catalyst

An enhancement of conventional transesterification is direct transesterification. In this method, to produced biodiesel, the biomass is reacted directly with the short-chain alcohol and the catalyst in one step. To overcome some drawbacks of the homogeneous direct transesterification reaction, heterogeneous catalysts have been used. The heterogeneous *in situ* or direct transesterification is expected to be an effective FAME production process with low cost and minimal environmental impact because of simplifying the production under mild reaction conditions. It is a cheaper method than conventional because the extraction of the triglycerides and the transesterification are done *in situ* at the same time decreasing the time to obtaining biodiesel and the excessive use of different resources. However, there are few studies were reported with solid heterogeneous catalyst in direct transesterification.

Direct transesterification has been reported using different heterogeneous catalysts. As acid heterogeneous catalyst, the CT 269 ion-exchange resin was reported to obtain biodiesel from microalgae [51]. The optimum conditions of this reaction were 95°C, and mass ratio catalyst/biomass equals to 0.52:1. On the other hand, the heterogeneous basic catalyst reported was LiOH-pumice [52] obtained by acid treatment and wet impregnation. The highest yield obtained was 47% with 20 wt% of catalyst at 80° C in 3 hours of reaction time and a relation methanol/biomass 12 mL/g. The biomass used in this reaction was *Chlorella sp. microalgae*. The same microalgae were used to produce biodiesel using carbon-dot functionalized strontium oxide [53]. The reaction was carried out with microwave radiation with dried microalgae mixed with chloroform, methanol, and 0.3 g of the catalyst. The temperature of the reaction was 60° C, the conversion of the lipids into FAME's was 97 wt%, and the maximum yield was 45.5%. CaO obtained from eggshell waste was evaluated in direct transesterification from *A. obliquus* microalgae. The eggshell needed a previous treatment to obtain the CaO as catalyst and obtained FAMEs. The biomass was mixed with the catalyst and methanol with a ratio 10:1 wt/vol; the temperature of the reaction was 70°C for a period of 1–5 h. About 86.41% was the yield reported using 1.7% (w/w) [54]. Biodiesel was produced from palm kernel by *in situ* transesterification using CaO as catalyst. The biomass was mixed with methanol and CaO. The reaction was carried out at 65° C for 3 hours. The oil content in the biomass was 33.08%. In this case, the size of the catalyst was very important, and the optimal size was <1 mm approximately [55]. In another study, strontium oxide as catalyst was evaluated using castor and jatropha seeds with microwave and ultrasound irradiation. The yield obtained with castor seeds was 57.2% of the total weight, and the conversion into FAMEs was 99.9%. With jatropha seeds, the yield was 41.1% with a conversion of the oil into FAMEs of 99.7% by microwave irradiation. By ultrasound irradiation, the yields were 48.2% and 32.9% from castor and jatropha seed, respectively [32].

Recently, Na_2ZrO_3 was evaluated as heterogeneous catalyst using *J. curcas L.* and *R. communis* seeds. This ceramic material exhibits interesting catalytic properties as basic catalysts and their low solubility and high stability. For heterogeneous direct transesterification, the seeds were ground and mixed with methanol and the catalyst with a molar ratio 1:65 (oil/methanol) at 65°C. At the first reactions, n-hexane was added to the reaction to favor the vegetable oil extraction. The presence of n-hexane in the reactions using 10 wt% of Na_2ZrO_3 as catalyst decreases the conversion of FFAs in 76% for *Jatropha Curcas L.* and 66% for *R. communis* seeds [21, 22]. This is caused by the decrement in the contact area between the reaction products caused by the non-polar solvent despite the increase to 10 wt% of the catalyst. The ideal conditions using only methanol as reactant and solvent were 5 wt% of catalyst at 65° C for 8 hours obtaining conversion to FAME of 99.9% in the first cycle and 72.5% after 5 cycles of reusing the catalyst. **Figure 7** shows the kinetic curves of the FAME yield of *in situ* transesterification reaction of Jatropha seeds. The use of methanol as reactant and solvent in transesterification reaction represents an environmental benefit due to methanol being environmentally friendly, and it can be a replacement for n-hexane or dichloromethane, which are toxic and hazard solvents [21].

As with jatropha seeds, the heterogeneous direct transesterification of *Ricinus communis* seed was performed with Na_2ZrO_3 in the presence of methanol. The yield obtained with optimal conditions (65° C, 5 wt%) was ~99.9% just in 4 hours. **Figure 8** discloses the FAME yield *versus* the reaction time. The selectivity between the catalysts

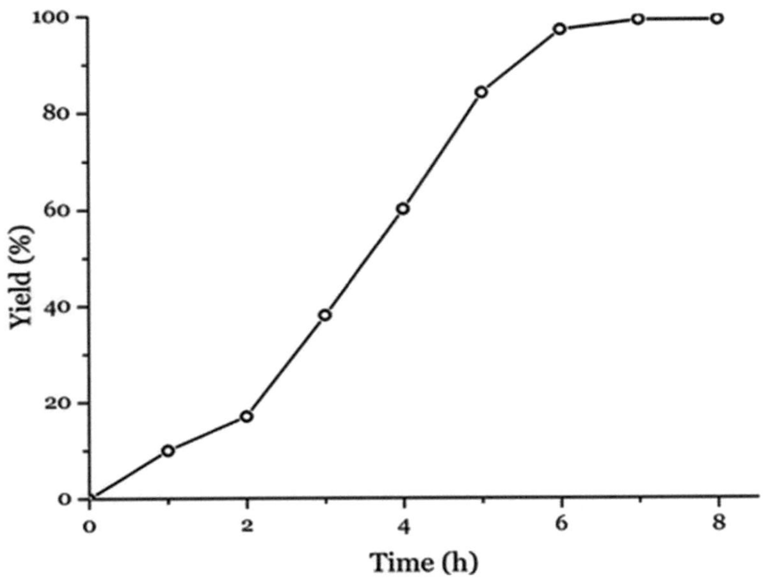

Figure 7.
Kinetic curves of the FAME conversion efficiency of in-situ *transesterification reaction of Jatropha seeds and methanol heterogeneous catalysts.*

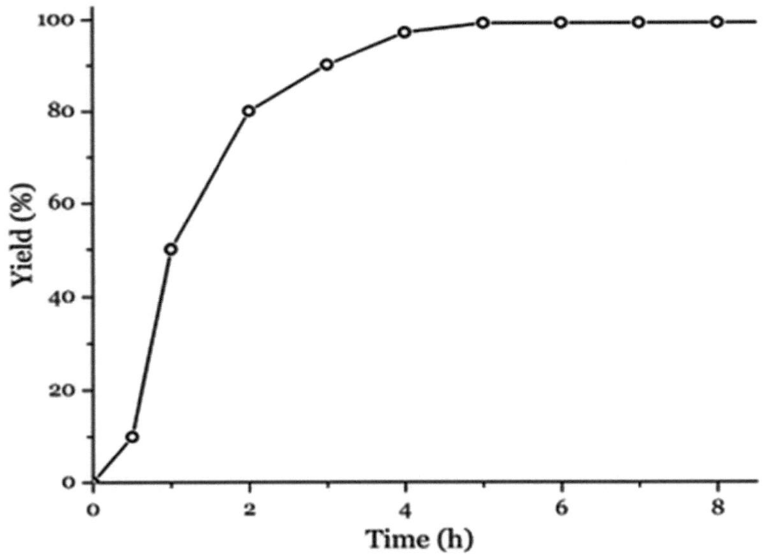

Figure 8.
kinetic curves of the FAME conversion efficiency of in situ *transesterification reaction of* Ricinus communis *seeds and methanol heterogeneous catalysts.*

reduced the reaction time and increased the yield and conversion. The hydroxyl group, -OH, of the ricinoleic methyl ester interacts with the high surface basicity of the solid catalyst, increasing their catalytic activity (**Figure 5B** shows the 1H-NMR spectra of the FAME *R. communis*, similar spectra). In this reaction *n*-hexane was substitute by methanol. The toxicity and environmental hazard decreased. Methanol as solvent increased the greenness of heterogeneous process, and the products obtained show high purity.

In addition, some reusing reactions are being carried out to determine the selectivity and stability of the catalyst. The heterogeneous direct transesterification (one-step process) using Na-based catalyst is a promising alternative for more sustainable, cleaner, and efficient FAME production.

After a few reusing reactions, a reduction in FAME conversions was observed. This may be attributed to the adsorption of organic matters from the biomass and the glycerol on the catalyst surface, which generates a low contact area between the oil and the Na_2ZrO_3.

In recent years, heterogeneous acid catalysts have been reported in biodiesel production and are an emerging field of research. But before producing competitive biodiesel, it is necessary to improve some aspects of catalyst as the stability of acid sites and the control of the surface properties.

3. Conclusions

Direct transesterification reaction from non-edible crops was successfully evaluated using both homogeneous and heterogeneous catalysts, from *Jatropha curcas L* and *R. communis* seeds, in one step. However, heterogeneous direct transesterification provides a new alternative and promising method to obtain high-purity biofuels. No toxic organic solvent was used to extract the oil from seeds, or no water was used to neutralize the products. Furthermore, methanol acts as reactant and solvent in the *in situ* reaction. This method improved the FAMEs production because the heterogeneous catalyst produces maximum conversion of ~99.9% with 5 wt% of catalyst at 65°C. For *R. communis*, the optimal conditions were reached in 4 hours of reaction time due to the -OH group. In addition, the reuse and the stability of the solid catalyst in the direct transesterification reactions with yields of>72.5% in fifth cycle were evaluated. However, it is necessary investigate the optimal condition to increase the yield of FAME by favoring the contact between the biomass and the catalyst. Finally, heterogeneous direct transesterification has potential environmental and energy benefits in comparison with the conventional biodiesel production method.

4. Transesterification future research direction

The processes for the biodiesel production must be carried out in a sustainable way using third- and fourth-generation feedstocks. Currently, our work group is focused on the development of green processes for the preparation of biodiesel through green routes that imply incorporating principles of sustainable chemistry. These clean processes incorporate the principles of a) atomic efficiency, b) use of renewable raw materials, c) preparation of heterogeneous catalysts, d) prevention of waste generation, and e) heterogeneous design reactors.

a. According to the results discussed in this chapter (Section 2), direct transesterification *(in situ)* is a promising alternative for obtaining biodiesel, this method simplifies the conventional process and eliminates the oil extraction stages. With this chemical synthesis, we are reducing the use of solvents and making the process of obtaining biodiesel more efficient and environmentally friendly in short times.

b. The use of raw materials or biomass considered as waste generates an added value to the transesterification process. For example, according to data from the United Nations Food and Agriculture Organization, more than 6.5 million tons of avocado are produced, where Mexico contributes 34% with approximately 2.1 million tons. In this research, the avocado stone is incorporated as a raw material to produce biodiesel, considered a renewable resource.

c. It is being proposed to prepare heterogeneous catalysts from new synthesis methods or using biomass waste as precursor, for example, eggshell waste because of the high content of CaO. There is great interest in these types of catalysts for environmental, energy, and economic reasons. On the other hand, studies are still being carried out on its reuse in the direct transesterification cycles. Also, it is important to investigate the interaction between seeds and the catalyst.

d. Biomass as residual by-product of direct transesterification reactions of the oil seeds is a value-added product. One proposal for its use is to obtain biopolymers; that is, these biopolymers are expected to have applications in the manufacture of bags, and that they can replace conventional plastics.

e. One of the technological challenges in biodiesel production is the design of heterogeneous catalysts. Thus, the scaling of heterogeneous processes is a key for the implementation of direct transesterification. The use of heterogeneous catalysts would result in simpler, cheaper separation processes as well as capital and energy costs.

Acknowledgements

Authors are grateful to Centro de Nanociencias y Micro y Nanotecnología del Instituto Politécnico Nacional (IPN) for their assistance in ESI-MS and 1H-NMR analyses. Also, the authors greatly appreciated the support from SECTEI CM-059/2021, SIP-IPN 2104 (modulo 20220625) projects and BEIFI scholarship. Finally, A. Martínez Ponce acknowledges the financial support from the DGAPA-UNAM PAPIME PE106522 project.

Author details

Issis Claudette Romero-Ibarra[1*], Araceli Martínez Ponce Escuela[2], Gabriela Elizabeth Mijangos Zúñiga[3] and Wendy Eridani Medina Muñoz[3]

1 National Polytecnical Institute (Instituto Politécnico Nacional, Unidad Profesional Interdisciplinaria en Ingeniería y Tecnologías Avanzadas - UPIITA), Mexico City, Mexico

2 Nacional de Estudios Superiores, Unidad Morelia, Universidad Nacional Autónoma de México, Morelia, Michoacán, Mexico City, Mexico

3 National Polytecnical Institute (Instituto Politécnico Nacional, UPIITA), Mexico City, Mexico

*Address all correspondence to: iromero@ipn.mx

References

[1] Kumar A, Sharma S. An evaluation of multipurpose oil seed crop for industrial uses (Jatropha curcas L.): A review. Industrial Crops and Products. 2008;**28**: 1-10. DOI: 10.1016/j.indcrop.2008.01.001

[2] Reddy B, Ramesh S, Kumar A, Wani SP, Ortiz R, Ceballos H, et al. Bio-fuel crops research for energy security and rural development in developing countries. Bioenergy Research. 2008;**1**: 248-258. DOI: 10.1007/s12155-008-9022-x

[3] Joshi G, Rawat DS, Sharma AK, Pandey JK. Microwave enhanced alcoholysis of non-edible (algal, jatropha and pongamia) oils using chemically activated egg shell derived CaO as heterogeneous catalyst. Bioresource Technology. 2016;**219**:487-492. DOI: 10.1016/j.biortech.2016.08.011

[4] Achten WMJ, Mathijs E, Verchot L, Singh VP, Aerts R, Muys B. Jatropha biodiesel fueling sustainability. Biofuels, Bioproducts and Biorefining. 2007;**1**: 283-291. DOI: 10.1002/bbb.39

[5] Hincapié G, Mondragón F, López D. Conventional and in situ transesterification of castor seed oil for biodiesel production. Fuel. 2011;**90**(4): 1618-1623. DOI: 10.1016/J.FUEL.2011.01.027

[6] Koh MY, Mohd GTI. A review of biodiesel production from Jatropha curcas L. oil. Renewable and Sustainable Energy Reviews. 2011;**15**:2240-2251. DOI: 10.1016/j.rser.2011.02.013

[7] Divakara BN, Upadhyaya HD, Wani SP, Laxmipathi CL. Biology and genetic improvement of Jatropha curcas L.: A review. Applied Energy. 2010;**87**: 732-742. DOI: 10.1016/j.apenergy.2009.07.013

[8] Kuete V. Physical, hematological, and histopathological signs of toxicity induced by African medicinal plants. In: Kuete V, editor. Toxicological Survey of African Medicinal Plants. Cameroon: Faculty of Science, University of Dschang; 2014. pp. 635-657. DOI: 10.1016/B978-0-12-800018-2.00022-4

[9] Melero JA, Iglesias J, Morales G. Heterogeneous acid catalysts for biodiesel production: Current status and future challenges. Green Chemistry. 2009;**11**:1285-1308. DOI: 10.1039/B902086A

[10] Luque R, Lovett JC, Datta B, Clancy J, Campeloa JM, Romero AA. Biodiesel as feasible petrol fuel replacement: Amultidisciplinary overview. Energy & Environmental Science. 2010;**3**:1706-1721. DOI: 10.1039/C0EE00085J

[11] Marx S. Glycerol-free biodiesel production through transesterification: A review. Fuel Processing Technology. 2016;**151**:139-147. DOI: 10.1016/j.fuproc.2016.05.033

[12] Tan HW, Abdul AR, Aroua MK. Glycerol production and its applications as a raw material: A review. Renewable and Sustainable Energy Reviews. 2013; **27**:118-127. DOI: 10.1016/j.rser.2013.06.035

[13] Atadashi IM, Aroua MK, Abdul AR, Sulaiman NMN. The effects of catalysts in biodiesel production: A review. Journal of Industrial and Engineering Chemistry. 2013;**19**(1):14-26. DOI: 10.1016/j.jiec.2012.07.009

[14] Lotero E, Liu Y, Lopez D, Suwannakarn K, Bruce D, Goodwin J. Synthesis of biodiesel via acid catalysis.

Industrial & Engineering Chemistry Research. 2005;**44**(14):5353-5363. DOI: 10.1021/ie049157g

[15] Lam MK, Lee KT, Mohamed AR. Homogeneous, heterogeneous, and enzymatic catalysis for transesterification of high free fatty acid oil (waste cooking oil) to biodiesel: A review. Biotechnology Advances. 2010; **28**(4):500-518. DOI: 10.1016/j.biotechadv.2010.03.002

[16] Lee DW, Park YM, Lee KY. Heterogeneous base catalysts for transesterification in biodiesel synthesis. Catal Surv from Asia. 2009; **13**(2):63-77. DOI: 10.1007/s10563-009-9068-6

[17] Wali Khan I, Naeem A, Farooq M, Ali Z, Saeed T, Perveen F, et al. Biodiesel production by valorizing waste non-edible wild olive oil using heterogeneous base catalyst: Process optimization and cost estimation. Fuel. 2022;**320**:123828. DOI: 10.1016/j.fuel.2022.123828

[18] Abdulkareem A, Nasir N. Biodiesel production from canola oil using TiO_2/CaO as a heterogenous catalyst. Journal of Advanced Research in Fluid. 2022;**93**: 125-137. DOI: 10.37934/arfmts.93.2.125137

[19] Iskandinata I, Taslim T, Bani O, Purba HLM. Potential of papaya seeds as a heterogenous catalyst in biodiesel synthesis. IOP Conference Series: Earth and Environmental Science. 2022; **912**(2):24-25. DOI: 10.1088/1755-1315/912/1/012022

[20] Santiago-Torres N, Romero-Ibarra IC, Pfeiffer H. Sodium zirconate (Na_2ZrO_3) as a catalyst in a soybean oil transesterification reaction for biodiesel production. Fuel Processing Technology. 2014;**120**: 34-39. DOI: 10.1016/j.fuproc.2013.11.018

[21] Martínez A, Mijangos G, Romero-Ibarra IC, Hernadez-Altamirano R, Mena V, Gutiérrez S. A novel green one-pot synthesis of biodiesel from Ricinus communis seeds by basic heterogeneous catalysis. Journal of Cleaner Production. 2018;**196**:340-349. DOI: 10.1016/j.jclepro.2018.05.241

[22] Martínez A, Mijangos G, Romero-Ibarra IC, Hernadez-Altamirano R, Mena V. In-situ transesterification of Jatropha curcas L. seeds using homogeneous and heterogeneous basic catalysts. Fuel. 2019;**235**:277-287. DOI: 10.1016/j.fuel.2018.07.082

[23] Torres-Rodríguez DA, Romero-Ibarra IC, Ibarra IA, Pfeiffer H. Biodiesel production from soybean and Jatropha oils using cesium impregnated sodium zirconate as a heterogeneous base catalyst. Renewable Energy. 2016;**93**: 323-331. DOI: 10.1016/j.renene.2016.02.061

[24] Mijangos G, Cuautli C, Romero-Ibarra IC, Vazquez-Arenas J, Santolalla-Vargas C, Santes V, et al. Experimental and theoretical analysis revealing the underlying chemistry accounting for the heterogeneous transesterification reaction in Na_2SiO_3 and Li_2SiO_3 catalysts. Renewable Energy. 2022;**184**: 845-856. DOI: 10.1016/j.renene.2021.11.090

[25] Rodríguez-Ramírez R, Romero-Ibarra I, Vazquez-Arenas J. Synthesis of sodium zincsilicate (Na_2ZnSiO_4) and heterogeneous catalysis towards biodiesel production via box-behnken design. Fuel. 2020;**280**:118-668. DOI: 10.1016/j.fuel.2020.118668

[26] Leung DYC, Wu X, Leung MKH. A review on biodiesel production using

catalysed transesterification. Applied Energy. 2017;**87**(4):1083-1095. DOI: 10.1016/j.apenergy.2009.10.006

[27] Ma FA, Hanna M. Biodiesel production: A review. Bioresource Technology. 1999;**70**(1):1-15. DOI: 10.1016/S0960-8524(99)00025-5

[28] Shuit SH, Lee KT, Kamaruddin AH, Yusup S. Reactive extraction and in situ esterification of Jatropha Curcas L. seeds for the production of biodiesel. Fuel. 2010;**89**(2):527-530. DOI: 10.1016/J.FUEL.2009.07.011

[29] Liu Y, Tu Q, Knothe G, Lu M. Direct transesterification of spent coffee grounds for biodiesel production. Fuel. 2017;**199**:157-161. DOI: 10.1016/j.fuel.2017.02.094

[30] Kartika IA, Yani M, Ariono D, Evon P, Rigal L. Biodiesel production from jatropha seeds: Solvent extraction and in situ transesterification in a single step. Fuel. 2013;**106**:111-117. DOI: 10.1016/j.fuel.2013.01.021

[31] Harrington KJ, D'Arcy-Evans C. Transesterification in situ of sunflower seed oil. Industrial and Engineering Chemistry Product Research and Development. 1985;**24**(2):314-318. DOI: 10.1021/i300018a027

[32] Koberg M, Cohen M, Ben-Amotz A, Gedanken A. Bio-diesel production directly from the microalgae biomass of Nannochloropsis by microwave and ultrasound radiation. Bioresource Technology. 2011;**102**(5):4265-4269. DOI: 10.1016/j.biortech.2010.12.004

[33] Johnson MB, Wen Z. Production of biodiesel fuel from the microalga Schizochytrium Limacinum by direct transesterification of algal biomass. Energy & Fuels. 2009;**23**(10):5179-5183. DOI: 10.1021/ef900704h

[34] Hidalgo P, Toro C, Ciudad G, Navia R. Advances in direct transesterification of microalgal biomass for biodiesel production. Reviews in Environmental Science and Biotechnology. 2013;**12**:179-199. DOI: 10.1007/s11157-013-9308-0

[35] Dias L, Missio R, Dias D. Antiquity, botany, origin and domestication of Jatropha Curcas (Euphorbiaceae), a plant species with potential for biodiesel production. Genetics and Molecular Research. 2012;**11**(3): 2719-2728. DOI: 10.4238/2012.June.25.6

[36] Doan LG. Ricin: Mechanism of toxicity, clinical manifestations, and vaccine development. A review. Journal of Toxicology and Clinical Toxicology. 2004;**42**(2):201-208. DOI: 10.1081/CLT-120030945

[37] Kyari MZ. Extraction and characterization of seed oil. International Agrophysics. 2008;**22**:139-142

[38] Sayyar S, Abidin ZZ, Yunus R, Muhammad A. Extraction of oil from Jatropha seeds-optimization and kinetics. American Journal of Applied Sciences. 2009;**6**(7):1390-1395. DOI: 10.3844/ajassp.2009.1390.1395

[39] Perdomo FA, Acosta-Osorio AA, Herrera G, Vasco-Leal JF, Mosquera-Artamonov JD, Millan-Malo B, et al. Physicochemical characterization of seven Mexican Ricinus Communis L. Seeds & oil contents. Biomass and Bioenergy. 2013;**48**:17-24. DOI: 10.1016/j.biombioe.2012.10.020

[40] Fenn JB, Mann M, Meng CK, Wong SF, Whitehouse CM. Electrospray ionization for mass spectrometry of large biomolecules. Science (80-.). 1989;**246**: 64-71

[41] Vyas AP, Verma JL, Subrahmanyam NA. Review on FAME production processes. Fuel. 2010;**89**(1): 1-9. DOI: 10.1016/j.fuel.2009.08.014

[42] Mangesh GK, Dalai AK. Waste cooking oil an economical source for biodiesel: A review. Industrial and Engineering Chemistry Research. 2006; **45**(9):2901-2913. DOI: 10.1021/IE0510526

[43] Fukuda H, Kondo A, Noda H. Biodiesel fuel production by transesterification of oils. Journal of Bioscience and Bioengineering. 2001; **92**(5):405-416. DOI: 10.1016/S1389-1723(01)80288-7

[44] Aransiola EF, Ojumu TV, Oyekola OO, Madzimbamuto TF, Ikhu-Omoregbe DIO. A review of current technology for biodiesel production: State of the art. Biomass and Bioenergy. 2014;**61**:276-297. DOI: 10.1016/j.biombioe.2013.11.014

[45] Zhang Y, Dubé M, McLean D, Kates M. Biodiesel production from waste cooking oil: 1. Process design and technological assessment. Bioresource Technology. 2003;**89**(1):1-16. DOI: 10.1016/S0960-8524(03)00040-3

[46] Kasim FH, Harvey AP, Zakaria R. Biodiesel production by in situ transesterification. Biofuels. 2010;**1**(2): 355-365. DOI: 10.4155/bfs.10.6

[47] Kildiran G, Yücel SÖ, Türkay S. In-situ alcoholysis of soybean oil. Journal of the American Oil Chemists' Society. 1996;**73**(2):225-228. DOI: 10.1007/bf02523899

[48] Patil PD, Deng S. Optimization of biodiesel production from edible and non-edible vegetable oils. Fuel. 2009; **88**(7):1302-1306. DOI: 10.1016/j.fuel.2009.01.016

[49] Talha NS, Sulaiman S. Overview of catalysts in biodiesel production. ARPN Journal of Engineering and Applied Sciences. 2016;**11**(1):439-442

[50] Singh A, Gaurav K. Advancement in catalysts for transesterification in the production of biodiesel: A review. Journal of Biochemical Technology. 2018;**7**(3):1148-1158

[51] Vicente G, Carrero A, Rodríguez R, del Peso GL. Heterogeneous-catalysed direct transformation of microalga biomass into biodiesel-grade FAMEs. Fuel. 2017;**200**:590-598. DOI: 10.1016/j.fuel.2017.04.006

[52] De Luna MDG, Doliente LMT, Ido AL, Chung TW. In situ transesterification of chlorella sp. microalgae using LiOH-pumice catalyst. Journal of Environmental Chemical Engineering. 2017;**5**(3):2830-2835. DOI: 10.1016/j.jece.2017.05.006

[53] Tangy A, Kumar V, Neel Pulidindi I, Kinel-Tahan Y, Yehoshua Y, Gedanken A. In situ transesterification of Chlorella vulgaris using carbon-dot functionalized strontium oxide as a heterogeneous catalyst under microwave irradiation. Energy and Fuels. 2016; **30**(12):10602-10610. DOI: 10.1021/acs.energyfuels.6b02519

[54] Pandit PR, Fulekar MH. Egg shell waste as heterogeneous nanocatalyst for biodiesel production: Optimized by response surface methodology. Journal of Environmental Management. 2017; **198**:319-329. DOI: 10.1016/j.jenvman.2017.04.100

[55] Tarigan JB, Prakoso HT, Siahaan D, Kaban J. Rapid biodiesel production from palm kernel through in situ transesterification reaction using CaO As catalyst. International Journal of Applied Chemistry. 2017;**13**(3):631-646

Chapter 3

Biotechnological Interventions for the Production of Glycerol-Free Biodiesel

Muhammad Saeed, Ghulam Mustafa, Faiz Ahmad Joyia, Aneela Shadab and Aqsa Parvaiz

Abstract

Advances in plant biotechnology and microbial genetics are speeding up because of the urgent need to provide a steady supply of resources. Growing cost of crude oil is having a negative impact on economies throughout the globe. Just biodiesel and bioethanol have been recognized as viable fossil fuel replacements. Chemical catalysis is primary way to synthesize biodiesel, besides enzymatic and microbial methods also play important role in biodiesel synthesis. These processes may play a significant part in the replacement of petroleum-based diesel in the future. The growth of sustainable, economically feasible biotechnological tools for the synthesis of biodiesel requires strong collaboration among several disciplines. In this age, lipases are the preferred enzymes for producing methyl esters (FAME), which are significant biological objects in biodiesel, from fatty acid esters (FAE) derived from fats and oils. It has also been shown that designed whole-cell microorganisms may directly produce FAE (MicroDiesel). The expensive cost of the biocatalyst continues to be a barrier to current enzymatic procedures, although advancements have recently been achieved, enabling the first synthetic enzymatic biodiesel synthesis. The fabrication of biodiesel which is enzymatic is primarily desirable due to the initial materials (waste frying oils, oils that were having high water content, etc.), where standard interesterification which is chemical is seldom applicable.

Keywords: biodiesel, glycerol, sugarcane, biochar, cyanobacteria

1. Introduction

As a result of climate change-causing greenhouse gas emissions, depleting oil reserves, and skyrocketing crude oil prices, the global community is increasingly turning to biological processes and green technologies to create technological compounds and alternative fuels from renewable resources. Advances in plant biotechnology and microbial genetics are speeding up because of the urgent need to provide a steady supply of resources. There is capability in biotechnology to manufacture a large number of these compounds from inexhaustible sources, at present virtually all the main bulk chemicals, with the exception of ethanol, are produced through the petrochemical approach [1]. Fermentation products typically have substantially smaller production quantities (less than one million tonnes per year) but higher pricing compared to petrochemical-based compounds [2]. However, bio-based processes offer the benefits of using renewable feedstock, producing environmentally friendly emissions, and operating at relatively moderate temperatures and pressures. Although biotechnology has shown its worth in the fabrication of fine chemicals like organic acids, vitamins, and medicinal compounds, there has been a rising push to optimise green technologies, which is slowly shifting the situation. For example, oil which is a fossil fuel much likely to run out first. Unrefined oil is a multi-component mixture that consists of around 50–95% hydrocarbons. Nearly all glasshouse gas releases are caused by the burning of fossil fuels. In addition to lowering pollution levels, cutting down on fossil fuel consumption would significantly decrease the quantity of carbon dioxide generated. Inquisitiveness in the maintainable generation of chemical raw materials and fuels has increased in light of the known scarcity of crude oil compared to the vast availability of biomass. The quantity of biofuel produced is massive and is anticipated to produce exponentially in the coming years. The growing cost of crude oil is having a negative impact on economies throughout the globe, especially those of the industrialised countries that rely heavily on the commodity. Over the last decade, consumers have seen a quadrupling in the price of essential fuels including natural gas, gasoline, and diesel. Even though the United States and Europe are a wide range of users of fossil fuels, these innovations will also help Asia's rapidly expanding economy. As a result, there has been a rise in the study of potential substitutes for fossil fuels. Just biodiesel and bioethanol, however, have been widely recognised as viable fossil fuel replacements. When customers shop around for ways to save their energy bills, innovative technologies and efficiency improvements become hot commodities. Due to the dramatic changes in the cost of crude oil in recent months, various major oil importers have been actively looking for alternatives. More than only the oil shortage and environmental impact, biofuel's significance cannot be overstated. Feedstock and fuels which are liquids derived from resources that can be renewed will allow us to tap into the vast, as-yet-untapped potential of agricultural and forestry waste products. Other significant arguments in favour of the biofuel alternative include its positive effect on worldwide climate, increased security related to the environment and economics, and practically better and improved goods. Environmental difficulty and political will are necessary, but they are not sufficient to usher in a sustainable age, which is nevertheless hampered by economic constraints. Until recently, biotechnology was seen by the chemical industries as an excessively costly high-tech technique that was not suitable for use on a commercial scale. The current research efforts of large chemical firms to create bulk compounds like 1,3-propanediol using biological approaches show that this perspective is gradually changing. Despite government support, bioethanol has become the biggest fermentation product in use today, demonstrating biotechnology's potential for large-scale chemical production.

2. Bioethanol from glycerol as raw material in biodiesel production

Most delegates to the UN Climate Change Conference in Doha, Qatar, in December 2012 committed to even deeper cuts in carbon dioxide emissions and established the latest objectives and restrictions to be executed under the Kyoto Protocol Extension (2012–2020). In addition, the price of a barrel of Brent oil has risen as high as 87.19 Euros in the last year (www.indexmundi.com). As alternatives to fossil fuels, biofuels like bioethanol and biodiesel also help the environment. As a result, governments in many developed and developing nations are establishing and expanding research and policy initiatives to boost biofuels' production and consumption. EU leaders have decided to prioritise biofuels in their quest to meet their renewable energy goal of 20% of total energy consumption by 2020. The European Commission has set a target of 10% of total fuel consumption in transportation by 2020. In contrast to fossil fuels, biodiesel may be replenished again and over again. It may be made using either animal or plant-based fats or oils. Since the inception of diesel engines in 1893, the potential of utilising vegetable oils as fuel has been recognised. Vegetable oil has the potential to be an operational substitute fuel oil, but its high viscosity prevents its usage in most conventional diesel engines. Several techniques exist for reducing the thickness of vegetable oils. Methods such as diluting, microemulsifying, pyrolysing, and transesterifying are used to lessen the thickness of a substance. Most biodiesel producers employ transesterification, a process that lowers oil viscosity. The chemical process by which oil is changed into its fatty ester is termed transesterification, although it is also known as alcoholysis. Triglycerides (1) are converted into ethyl esters of fatty acids (3) and glycerol (4) by the transesterification process, in which alcohol (ethanol or methanol) (2) is reacted with a catalytic base (1), (4) (**Figure 1**).

Adding a catalyst speeds up and increases the yield of a reaction. Since the reaction might go either way, an excess of alcohol is utilised to tip the scales in favour of the product side of the equilibrium. An oil-splitting catalyst like NaOH or KOH and an alcohol-like methanol or ethanol are needed for the biodiesel process. Glycerol is the primary output. You may recognise this kind of crude glycerine by its dark colour and syrupy consistency. In contrast to petro-diesel, biodiesel is superior in many ways: flash point, sulphur content, biodegradability, and aromatic content [3]. Transesterification of vegetable oils typically employs homogeneous catalysts. Under moderate reaction conditions, base homogeneous catalysts like NaOH and KOH are the most effective. However, the reaction time for using acid homogenous catalysts is much higher. On the other hand, the price of the production of biodiesel is inflated by the need of treating the waste produced by homogeneous catalysts, which is itself difficult to recover. Biodiesel may be made from vegetable oils, and heterogeneous catalysts are a viable option for this process. Researchers are looking at the transesterification activity of several heterogeneous catalysts [4]. Heterogeneous catalysts have many advantages over their homogeneous counterparts, including the fact that they can be reused multiple times without degrading their performance and the ability to use an uninterrupted procedure without the need for additional decontamination steps, not to mention the possibility of being relatively inexpensive.

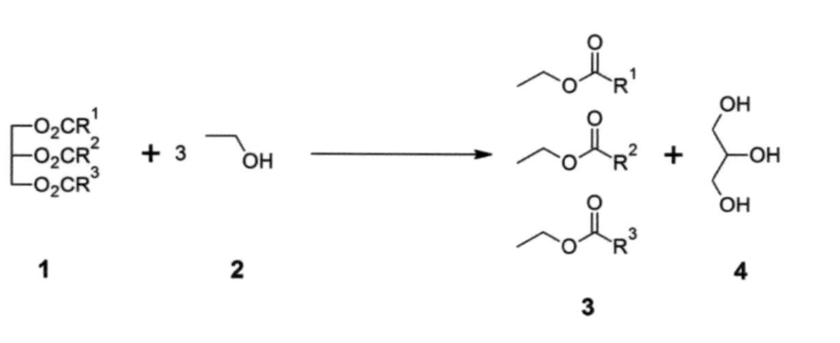

Figure 1.
Transesterification of triglycerides with alcohol.

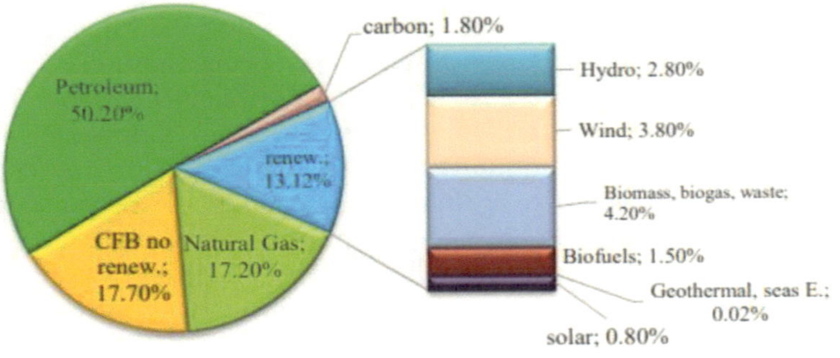

Figure 2.
The total energy used in 2010 (REP 2011/2022). The US and Brazilian markets account for the bulk of bioethanol sales, but biodiesel use has surged in the EU. At the end of 2010, Spain has more than 47 biodiesel facilities and four bioethanol plants with a combined installed production capacity of over 4 million tep (Per 2011/2020). The biofuel industry is experiencing a period of the technical revolution that is largely influencing the range of materials amenable to use in production processes. As such, the European Industrial Initiative on Bioenergy's implementation strategy prioritises thermochemical and biochemical processes for material conversion. Rapid expansion is predicted for the biodiesel industry. Its popularity will rise steadily over the next decade, helped by new regulations requiring the labelling of some mixtures. From 2011 to 2020, bioethanol use is predicted to roughly treble. The anticipated increase is due in part to the elimination of gas price subsidies and the standardisation of the labelling of gasoline blends. However, fresh environmental and economic questions regarding the feasibility of alternative fuels have been raised due to the increasing demand for biofuel, in particular biodiesel.

Financial incentives for both producers and consumers are still needed to increase biofuel usage. In this way, many governments subsidise the biofuel industry via tax credits and other measures. In the beginning, the EU subsidised biofuels like bioethanol and biodiesel. The United States government offers tax breaks to domestic producers in the amount of $0.51 per gallon for bioethanol and $1.00 per gallon for biodiesel. Presently, Germany incentivizes the use of biodiesel by taxing it at a lower rate than regular fuel. The yearly fabrication and use of biodiesel in Germany exceeds 2.5 billion litres [5]. The usage of biodiesel has spread to other EU nations, often as an additive to petroleum fuel. However, using objectives to stimulate the future production and use of biofuels are significantly more essential than tariffs and subsidies. Spain's Renewable Energy Plan (REP) 2011–2020 sets goals in conformity with the European Parliament's Directive 2009/28/EC on the encouragement of the utilisation of energy from sources which can be renewed. The goal of the REP is to meet the target set by the EU Directive, which calls for the use of renewable sources to account for at least 20% of gross final energy exhaustion by the year 2020. The International Energy Agency (IEA) reports that in 2010, biofuels substituted for 2% of global oil production. The amazing expansion of renewable energy sources over the last several years may be attributed in large part to the government assistance provided under the Renewable Energy Plan 2005–2010 (**Figure 2**).

3. Glycerol, from major commodity to waste effluent

Pure 1,2,3-propanediol (glycerol) is a colourless, odourless, hygroscopic, viscous liquid having the chemical formula OCH2CHOHCH2OH. It is a member of the alcohol family of organic compounds. The glycerol molecule's space-filling model is seen in **Figure 3**.

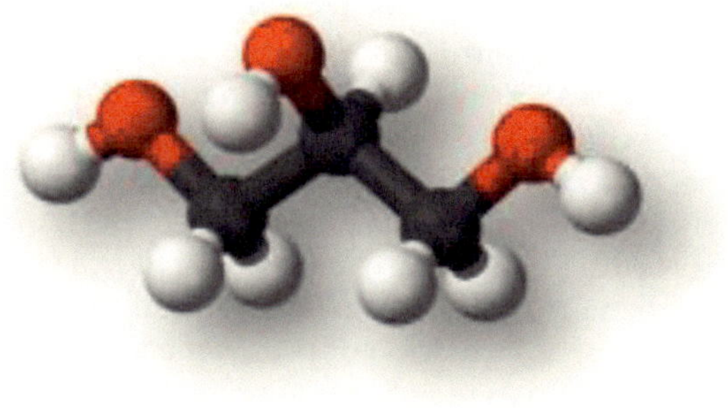

Figure 3.
Space-filling model for glycerol molecule.

Glycerol interacts with inorganic and organic acids to produce ethers, aldehydes, esters, and numerous derived molecules because it has one secondary and two primary alcohol groups per molecule. When there are several alcohol groups present, it is easier to make polymers and coatings (polyesters, polyether, and alkyd resins). Glycerol is a versatile chemical that may be used in a wide variety of industries due to its solubility in water, its biocompatibility, its lack of toxicity when applied topically, and its legal status in the food and drug industries. Also, glycerol's non-toxic qualities enable it for a wide variety of applications. It is safe to say that the cosmetics, explosives, food, pharmaceutical, polymer, and printing sectors are not the only ones that put this chemical component to good use (**Figure 4**). In the construction, automotive, and textile sectors, it is used to make gums and resins. In addition to its uses as a stabiliser in ice cream and a softening agent in baked products, mono and diglyceride emulsifiers also include this substance. As an added bonus, glycerol has several applications in the medical and pharmaceutical fields. It has also been put to use as a safe medium for freezing biological cells, which is a relatively new yet crucial.

Chemical synthesis from petrochemical feedstock or microbial fermentation [7] is both viable methods for glycerol production. In the last 150 years, scientists have learned that glycerol may be made by microorganisms. During World War I, glycerol was manufactured in industries by microbial fermentation. Due to poor glycerol yields and the difficulties of extracting and purifying glycerol from broth, microbial

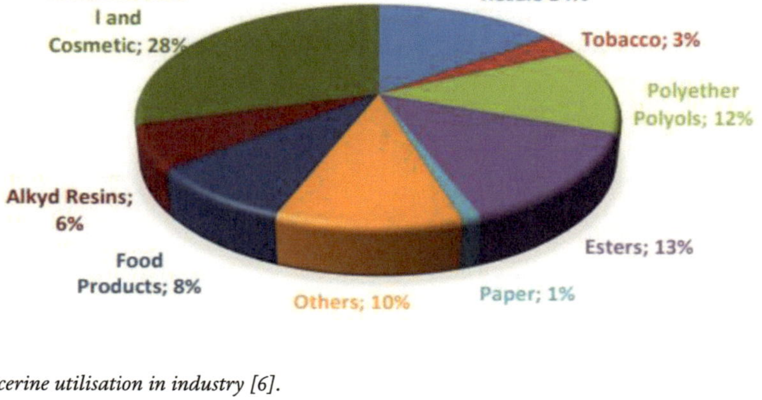

Figure 4.
Principal glycerine utilisation in industry [6].

synthesis eventually fell in favour of chemical synthesis from petrochemical feedstocks. At present, roughly 600,000 tonnes of glycerol are generated each year, and most of this comes from the saponification of lipids as a byproduct of soapmaking (**Figure 5**) [7]. The widespread use of detergents in industrialised countries has reduced the relevance of this step [8].

There are a number of ways to convert propylene into glycerol. Chlorinating propylene yields allyl chloride, which is oxidised with hypochlorite to form dichlorohydrin, which interacts with a robust base to yield epichlorohydrin, the most essential step in the epichlorohydrin process. Glycerol is produced once epichlorohydrin is hydrolysed. The oxidation or chlorination of propylene accounted for around 25% of 2001 global glycerol production in the chemical sector [7]. However, environmental concerns and the rising cost of propylene have caused this route's popularity to decline. Glycerol is a byproduct of biodiesel production, therefore the rising popularity of this alternative fuel has led to a slump in the glycerol market and rendered the epichlorohydrin technique for glycerol synthesis unprofitable on a commercial scale. By 2020, it is expected that there would be a glycerol surplus of six times the yearly need. Glycerol, on the other hand, might be used as a substrate in emerging commercial fermentation processes. Bioconversion of glycerol has resulted in a wide variety of useful byproducts. chotIn particular, the microbial synthesis of 1,3-propanediol from glycerol has received a lot of attention because of the diol's various potential uses in the creation of novel polymers. Only around 62–85% (w/w) of the crude stream is glycerol [9, 10]. Fats (soaps), water, methanol (often used in Europe) or ethanol (typically used in the United States) and catalyst leftovers make up the rest (salts). In addition to carbon, hydrogen, and oxygen, raw glycerol also includes trace amounts of elements including calcium, magnesium, phosphorus, and sulphur [10]. Producers of biodiesel employ a variety of feedstocks, each of which contributes a unique range of purity values. Raw glycerol levels were observed to be greatest (76.6%) in waste vegetable oil, with lower levels (62%) being produced by mustard, rapeseed, canola, crambe, and soybean. Raw glycerol is mostly accumulated by biodiesel factories, which create 10 g of glycerol for 100 g of biodiesel produced [11]. The concentration of raw glycerol is also attributable to the bioethanol sector. Common industrial methods provide 4 g of glycerol for every 48 g of ethanol [12]. Pure glycerol quantities are quite large and growing. Glycerol costs around $1200 per tonne in 2003. In 2006, the price per tonne was approximately $600 and declining [13]. Glycerine spot prices in Europe ranged from €260 to €350 per tonne in the month of November 2009. Even crude glycerol is no longer worth what it used to. US$0–$70 per tonne was the stated price range in 2006 [13]. Raw glycerol is worthless to those who manufacture it on a small basis. In these situations, the crude stream is often transported to a purification plant at the expense of the producers, disposed of in a landfill, or held in containers

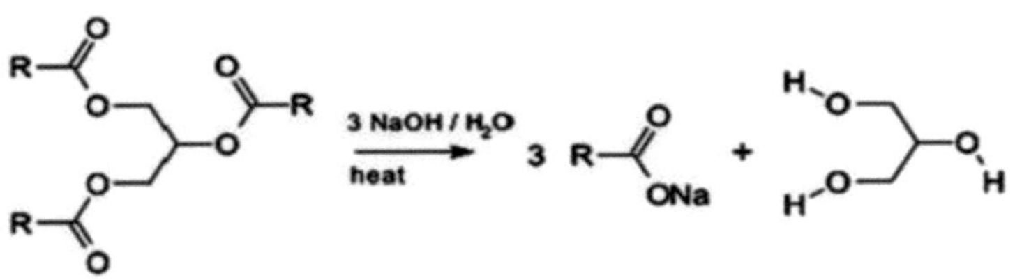

Figure 5.
Saponification of fats.

until a solution is sought. Crude glycerine is currently considered a contaminant that must be disposed of at a cost since it is unfit for most glycerol markets. The dumping of polluted waste glycerol from biodiesel production has caused environmental issues and driven down global glycerol market prices.

4. Method and approaches for glycerol-free biodiesel production

4.1 Emerging trends of microorganism in the biodiesel production

A variety of feedstocks (such as lipids, starch, and sugars) that may be utilised for the generation of biofuels are harvested and processed via the photosynthetic activities of plants (including cyanobacteria, algae, trees, grasses, and crops). Biofuels such as methane, ethanol, and diesel are already widely produced using already accepted "first generation" biofuel systems based on crop plants like sugar beet (*Beta vulgaris*), rapeseed (*Brassica napus*), sugarcane (*Saccharum* spp.), wheat (*Triticum* spp.), oil palm (*Elaeis oleifera*), soya beans (*Glycine max*), and corn (*Zea mays*). Pressure on food supply has resulted in increased worry and has sparked a heated "food versus fuel" debate, as a consequence of an expanding global population and significant droughts in key grain exporting countries (such as Australia). So, scientists are working on a new generation of land-free biofuel technologies. Most significantly, lignocellulosic methods are being developed to transform plant-based cellulose materials into liquid fuels. The most promising "non-food" plant possibilities for these methods include sorghum, miscanthus, camelina, switchgrass (*Panicum virgatum*), and poplar trees (*Populus* spp.). However, these systems can only be effective if scientists discover and implement energy-saving production techniques, such as enzymatic lignin digesting procedures (although chemical digestion techniques are also being investigated). Even if the resulting need for enzymes seems like a manageable obstacle, this technique may eventually add to food versus fuel dilemmas because of the existing reliance on appropriate land, most of which is already forested. Unless only waste products from existing agricultural or forestry systems are utilised, or feedstocks grown on non-arable land can be created, this might lead to a forest versus fuel dilemma.

The effective synthesis of starch, sugars, and oils by many microalgae makes them ideal feedstocks for the manufacture of biofuels such as biodiesel, ethanol, butanol, methane, and hydrogen. These microalgae may be cultivated in salt water. Microalgae may help with carbon capture because they take up carbon dioxide (CO_2) during growth from the atmosphere and, in certain circumstances, from industrial sources. To store carbon, the leftover waste biomass from fuel production may be pyrolysed to create a charcoal-like product (Biochar) that is stable over time. Biochar may replace coal as a fuel source, or it can be sold to the public as a soil amendment. While in principle microalgal biofuel systems may solve both the food versus fuel and the prospective forest versus fuel issues, no such system has yet reached commercial viability. In spite of widespread and overwhelming excitement, which Emily Waltz calls "algae ardour," there are firms actively pushing these technologies towards commercial operation, as Waltz revealed recently. Investment in microalgal biofuels has actually increased after the first failure of a start-up rather than decreasing. This investment makes sense in light of recent economic case studies on standalone microalgal biofuel production models and on a model that co-produces high-value products (HVPs). Key economic determinants, such as building costs, biomass productivity, and cost of the dominating output, were found via sensitivity analysis in these models (and its production in the case of high-value co-products) (**Figure 6**).

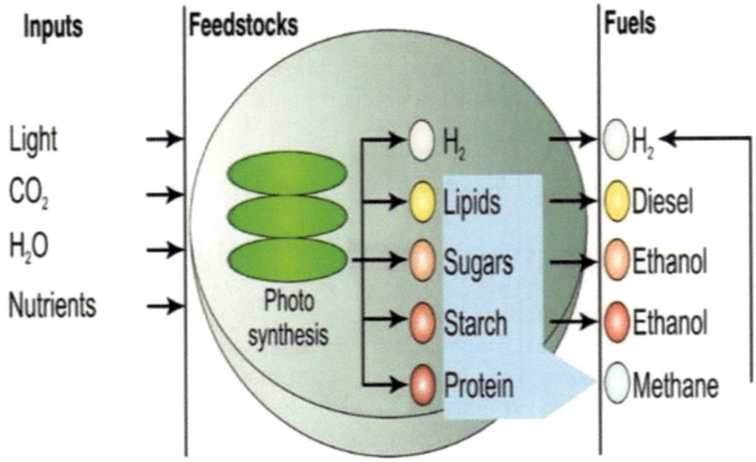

Figure 6.
Conversion of raw material into ethanol by photosynthesis.

4.2 Biotechnological production of biodiesel fuel using biocatalysed transesterification

Under supercritical circumstances, Saka and Kusdiana found that methanol facilitated the transesterification of triglyceride (TG) into fatty acid methyl ester (FAME). It is well-known that three sequential reversible processes may be used to convert methanol into biodiesel. First, triglycerides are broken down into smaller molecules called diglycerides (DGs), and then the DGs are broken down further into monoglycerides (MGs). The last process is the transformation of MGs into glycerol. In every stage of the reaction, FAME is the end product. Transesterification results in the formation of three FAMEs, as seen in. The biodiesel generation in supercritical MTBE process is quite similar, consisting of three successive reversible reactions. Triglyceride combines with MTBE to form mono tert-butyl ether (DGE), and DGE then undergoes further reactions to form monoglyceride di tert-butyl ether (MGE). The end result of MGE reacting with MTBE is the production of FAME GTBE. Without the need for a catalyst, MTBE can transform TG into FAME. Although the supercritical MTBE approach operates at a higher temperature, its FAME yield is lower than that of the supercritical methanol route under identical reaction conditions. When utilising supercritical MTBE, however, the yield of FAME reaches 95.4 wt% after a 12-minute residence period, which is almost identical to that achieved using the supercritical methanol approach. Surprisingly, the supercritical MTBE approach has a larger FAME yield than the supercritical methyl acetate route. Since MTBE is less polar than methyl acetate, it is more reactive. Thus, at ambient temperature and pressure, MTBE is more miscible with oil than methyl acetate. In this case, MTBE's miscibility helps it get around the mass transfer issue.

4.3 Glycerol-free biodiesel using methyl acetate as acyl acceptor with bio-enzymes

Using enzymes as the biocatalyst might potentially reduce downstream separation costs since they eliminate the requirement for a separate solvent. Novozyme 435 has

DOI: http://dx.doi.org/10.5772/intechopen.108895

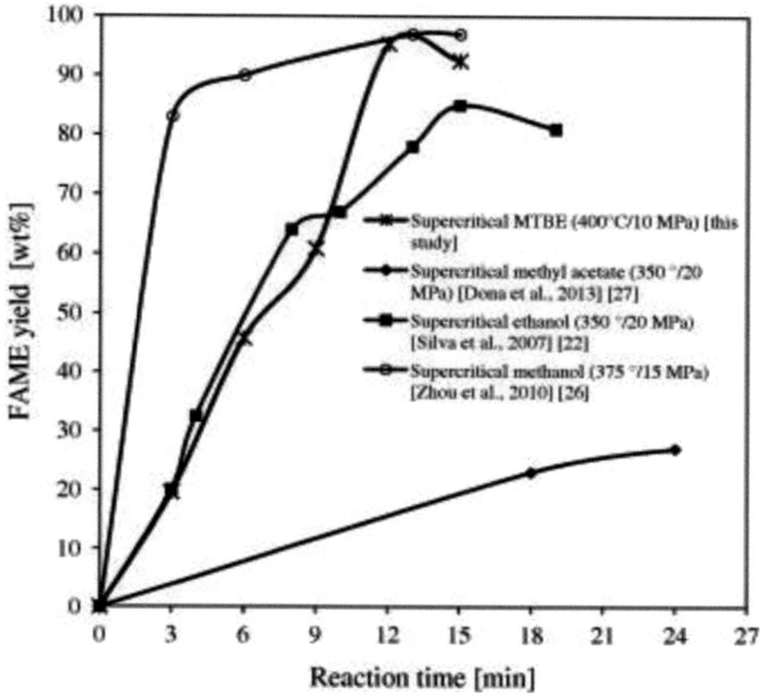

Figure 7.
The use of dimethyl carbonate to catalyse the enzymatic transesterification of vegetable oil.

been used in almost all published reports of enzymes for synthesis utilising dimethyl carbonate as an acyl acceptor, either as a free enzyme in solution or inactive on a poly-acrylic macro-porous substrate. Novozymes 435 includes lipase B from *Candida antarctica*, which is an approximate lipase. The summary of recently published research on the use of dimethyl carbonate to catalyse the enzymatic transesterification of vegetable oil is given here (**Figure 7**, **Table 1**).

4.4 Glycerol-free biodiesel using dimethyl carbonate as acyl acceptor with bio-enzymes

The environmental friendliness, safe for industrial use (low pressure and temperature is needed), low cost of feedstock (glycerol or glucose), and high theoretical molar yield of the biological conversion of glycerol to 1,3-propanediol are making it a more viable option than the chemical production approach. The sustainability of 1,3-propanediol's biotechnological production is enhanced by the fact that it may be made from a renewable resource. Although the fermentation by bacteria that converts glycerol to 1,3-propanediol has been investigated for about 120 years, its biotechnological potential has just been recognised and further study has only begun after 1990 [2]. Multiple microbial families have been documented as producing 1,3-propanediol since 1996. *Klebsiella*, *Enterobacter* [14], *Clostridium butyricum*, and *Klebsiella pneumoniae* stand out as the most promising producers due to their high substrate tolerance, abundant output, and high rates of production. Dual routes ferment glycerol through a dismutation mechanism. One of these routes involves the enzyme glycerol dehydrogenase converting glycerol to dihydroxyacetone, which then

Feedstock	Catalyst	Catalyst loading (g/g oil)	Temperature (°C)	Molar ratio (Oil:DMC)	Time (min)	Yield (wt.% diesel/oil)
Soybean oil	Various	5%	90°C	1:3.7	300	N 99.5
Refined soybean oil	$KOCH_3$	5%	90°C	1:9	300	7.6
			200°C (10 bar)		60	95.8
Refined canola oil	Triazabicyclo[4.4.0] dec-5-ene	2.5%	60°C	1:3	360	100%
Cotton seed oil	Lipase (Novozymes 435)	15% in petroleum ether	40°C	1:4.5	1440	~93
Waste cooking oil	Immobilised novozymes 435	10%	60°C	1:6	120	~87
Algae oil extracted from *Chlorella sp.*	Novozymes 435	50% in DMC	60°C	n/a	360	75.5
Waste cooking oil	Immobilised novozymes 435	20%	40°C	1:6	200	86.61
Purified soybean oil	Immobilised novozymes 435	20%	60°C	1:10	2880	96.4
Purified soybean oil	Immobilised novozymes 435	100 g/L in *t*-butanol	60°C	1:6	2880	84.9
Corn oil	Novozymes 435	10%	60°C	1:10	900	94
Palm oil	Immobilised novozymes 435	20%	55°C	1:10	1440	90.5
Chlorella sp. KR-1	Novozymes 435	20%	70°C	1:10	1440	36.7

Table 1.
Recent research on dimethyl carbonate as an acyl acceptor in non-supercritical biodiesel manufacturing.

goes through conventional glycolysis to generate pyruvate and may be further broken down into a variety of compounds such as acids and alcohols. The second process requires the reduction of 3-hydroxypropionaldehyde (3-HPA) to 1,3-propanediol through the expenditure of decreasing power NADH2 and the catalytic action of 1,3-propanediol:NAD oxidoreductase. This reduction requires coenzyme B12. Glycerol dehydratase is a critical limiting enzyme in the second metabolic pathway, which converts glycerol to 1,3-propanediol and is essential for cellular redox homeostasis. Theoretically, glycerol's anaerobic fermentation yields its greatest potential when acetate is the sole by-product [15–17]. The creation of waste substances not only limits the quantity of carbon accessible, but may also stifle the development of microbes.

Since acetic acid is necessary for the NADH2 process, the generation of 1,3-propanediol is reduced in tandem with the formation of all these other byproducts, especially ethanol and butanol. The bioconversion of glycerol to 1,3-propanediol has gained popularity since glycerol might sometimes be a surplus product. In the past 10 years, a lot of work has gone into making this procedure more efficient and raising the yield response. Research into genetically engineered strains for 1,3-propanediol synthesis has been conducted in an effort to boost the efficiency of naturally occurring producers during glycerol fermentations. Another issue with genetically modified microorganisms is that it is difficult to conduct fermentations on an industrial level with these strains since they are so sensitive and fragile. If petroleum supplies are depleted, biotechnology might be the key to continuing 1,3-propanediol manufacturing [18].

4.5 Bioconversion of glycerol by *C. butyricum*

The age of *Clostridium*, a kind of bacterium, is estimated to be about 2700 million years. This organism predated Earth's oxygen atmosphere, which is why *Clostridium* species are oxygen-sensitive. For almost 60 years, scientists have known that anaerobic bacteria, including *Clostridium*, may ferment glycerol into 1,3-propanediol. All member microorganisms in this genus have rod-like shape that is Gram-positive, moderately big, heterotrophic, endospore producing, and motile. Some are psychotropic or thermopholic, but mesophilia is the norm for the vast majority. *Clostridium* thrives in anaerobic settings rich in organic resources. Clostridia bacteria are present everywhere in the environment and may be discovered in a diversity of habitats, including soils, feed, aquatic sediments, and the digestive systems of humans and animals. They can remain alive for extended periods of time in hostile environments because of their spore-forming abilities. As long as they are kept in a medium that provides them with food and the right temperature and humidity, a single bacterial strain may proliferate. Among the roughly 100 species that make up the genus *Clostridium* are both common, free-living bacteria, and serious diseases. *Clostridium* produces a wide variety of extracellular enzymes, which contribute to their robust metabolic activity. Sugars may be fermented by

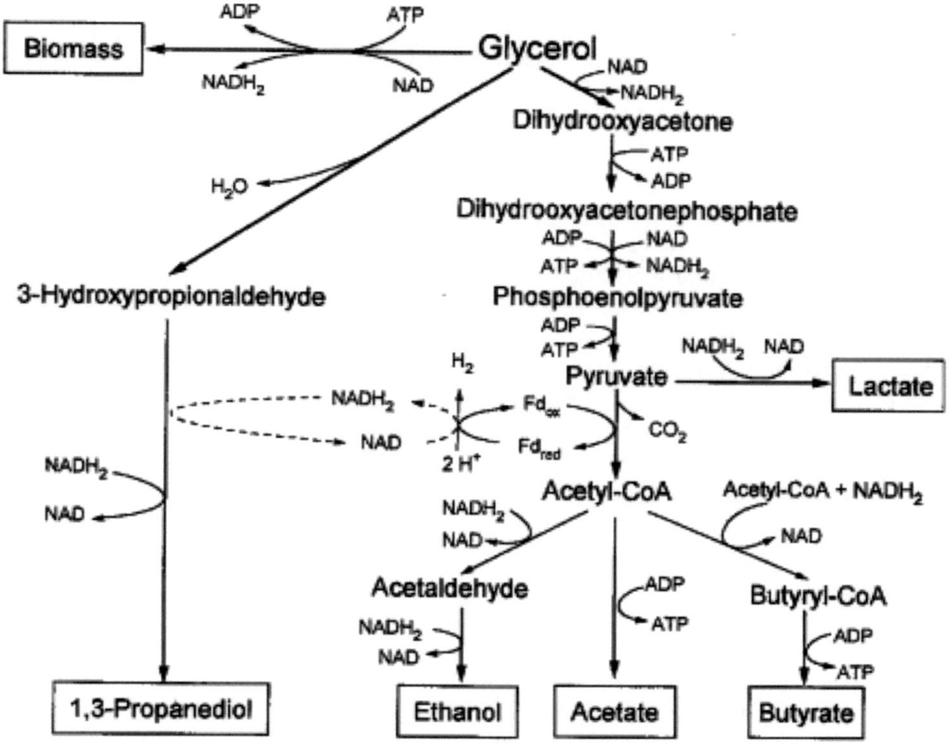

Figure 8.
Pathway of conversion of biomass into biodiesel.

these bacteria, leading to the production of hydrogen and organic molecules such as organic acids (particularly acetic acids and butyric), acetone, and butanol. Clostridium produces offensive-smelling breakdown products during the metabolism of amino and fatty acids [19]. Additionally, Clostridia can break down a variety of hazardous compounds and create chiral products, both of which need much effort to get by chemical synthesis. There are several distinct enzymes for breaking down starch and hemicellulose, and they have been found in many non-identical types of bacteria. *Clostridium thermocellum* is a prototypical cellulolytic *Clostridium*, producing a multienzyme cellulase complex that can break down cellulose, hemicellulose, and starch [20]. Using clostridial toxins and spores to treat human illness is a great advancement. Dystonias, involuntary muscular problems, pain, and other neurological conditions are treated with botulinum neurotoxin. Therapeutics are being developed to be delivered to tumours using Clostridia spore systems [21]. Clostridia have very basic nutritional needs. Typically, a complex nitrogen supply is necessary for optimal growth and solvent synthesis. There is a good deal of potential commercial interest in the non-pathogenic Clostridia. *C. butyricum* produces 1,3-PD and other byproducts during the fermentation of glycerol. *C. butyricum*'s anaerobic fermentation metabolic pathways produced by-products. These byproducts have a negative impact on *C. butyricum* development because they deplete the available carbon source (**Figure 8**).

5. Conclusion

Advances in plant biotechnology and microbial genetics are speeding up because of the urgent need to provide a steady supply of resources. The growing cost of crude oil is having a negative impact on economies throughout the globe. Cutting down on fossil fuel consumption would significantly decrease the quantity of carbon dioxide generated. Just biodiesel and bioethanol have been recognised as viable fossil fuel replacements. Biofuel's significance cannot be overstated.

Feedstock and fuels which are liquids derived from resources that can be renewed will allow us to tap into the vast, as-yet-untapped potential of agricultural and forestry waste products. Bioethanol has become the biggest fermentation product in use today, demonstrating biotechnology's potential for large-scale chemical production. Environmental difficulty and political will are necessary but not sufficient to usher in a sustainable age, hampered by economic constraints. The price of a barrel of Brent oil has risen as high as 87.19 Euros in the last year. Biofuels like bioethanol and biodiesel also help the environment. EU leaders have decided to prioritise biofuels in their quest to meet their renewable energy goal of 20% of total energy consumption by 2020. Biodiesel is superior to petro-diesel in many ways: flash point, sulphur content, biodegradability, and aromatic content. Transesterification of vegetable oils typically employs homogeneous catalysts, but heterogeneous ones are a viable option for this process.

Financial incentives for both producers and consumers are still needed to increase biofuel usage. The yearly fabrication and use of biodiesel in Germany exceed 2.5 billion litres. The biofuel industry is experiencing a period of technical revolution that is largely influencing the range of materials amenable to use in production processes. The US and Brazilian markets account for the bulk of bioethanol sales, but biodiesel use has surged in the EU. Biodiesel use is predicted to rise steadily over the next decade, helped by new regulations requiring the labelling of some mixtures.

Glycerol is a colourless, odourless, hygroscopic, and viscous liquid with the chemical formula OCH2CHOHCH2OH. It is a versatile chemical that may be used in a wide variety of industries due to its solubility in water and its lack of toxicity when applied topically. In the construction, automotive, and textile sectors, it is used to make gums and resins.

Author details

Muhammad Saeed[1], Ghulam Mustafa[1], Faiz Ahmad Joyia[1], Aneela Shadab[1] and Aqsa Parvaiz[2*]

1 Centre of Agricultural Biochemistry and Biotechnology, University of Agriculture, Faisalabad, Pakistan

2 Department of Agricultural Biochemistry and Biotechnology, The Women University Multan, Multan, Pakistan

*Address all correspondence to: aqsapervaiz333@gmail.com

References

[1] Wilke D. Chemicals from biotechnology: Molecular plant genetics will challenge the chemical and the fermentation industry. Applied Microbiology and Biotechnology. 1999;**52**:135-145

[2] Zeng A-P, Biebl H. Bulk chemicals from biotechnology: The case of 1,3-propanediol production and the new trends. Advances in Biochemical Engineering/ Biotechnology. 2002;**74**:239-259

[3] Bala BK. Studies on biodiesels from transformation of vegetable oils for diesel engines. Energy Education Science and Technology. 2005;**15**:1-45

[4] Yan S, DiMaggio C, Mohan S, Kim M, Salley SO, Simon Ng KY. Advancements in heterogeneous catalysis for biodiesel synthesis. Topics in Catalysis. 2010;**53**:721-736

[5] Silva GP, Mack M, Contiero J. Glycerol: A promising and abundant carbon source for industrial microbiology. Biotechnology Advances. 2009;**27**(1):30-39

[6] Mota C, Silva C, Gonçalves V. Gliceroquímica: Novos produtos e processos a partir da glicerina de produção de biodiesel. Quimica Nova - QUIM NOVA. 2009;**32**. DOI: 10.1590/ S0100-40422009000300008

[7] Wang Z-X, Zhuge J, Fang H, Prior BA. Glycerol production by microbial fermentation: A review. Biotechnology Advances. 2001;**19**:201-223

[8] Agarwal GP. Glycerol. Advances in Biochemical Engineering/Biotechnology. 1990;**41**:95-128

[9] Mu Y, Xiu ZL, Zhang DJ. A combined bioprocess of biodiesel production by lipase with microbial production of 1,3-propanediol by *Klebsiella pneumoniae*. Biochemical Engineering Journal. 2008;**40**:537-541

[10] Thompson JC, He BB. Characterization of crude glycerol from biodiesel production from multiple feedstocks. Applied Engineering in Agriculture. 2006;**22**(2):261-265

[11] Barbirato F, Himmi EH, Conte T, Bories A. 1,3-propanediol production by fermentation: An interesting way to valorize glycerin from the ester and ethanol industries. Industrial Crops and Products. 1998;**7**:281-289

[12] Yazdani S, Gonzalez R. Anaerobic fermentation of glycerol: A path to economic viability for the biofuels industry. Biotechnology. 2007;**18**:213-219

[13] Miller-Klein Associates. Impact of biodiesel production on the glycerol market. 2006. Available from: www.hgca.com

[14] Zhu M, Lawman PD, Cameron DC. Improving 1, 3-propanediol production from glycerol in a metabolically engineered *Escherichia coli* by reducing accumulation of snglycerol-3-phosphate. Biotechnology Progress. 2002;**18**:694-699

[15] Chotani G, Dodge T, Hsu A, Kumar M, LaDuca R, Trimbur D. The commercial production of chemicals using pathway engineering. Biochimica et Biophysica Acta. 2000;**1543**:434-455

[16] Saxena RK, Anand P, Saran S, Isar J. Microbial production of 1,3-propanediol: Recent developments and emerging opportunities. Biotechnology Advances. 2009;**27**:895-913

[17] Zeng A-P. Pathway and kinetic analysis of 1,3-propanediol production

from glycerol fermentation by *Clostridium butyricum*. Bioprocess Engineering. 1996;**14**:169-175

[18] Cho M-H, Joen SI, Pyo S-H, Mun S, Kim J-H. A novel separation and purification process for 1,3-propanediol. Process Biochemistry. 2006;**41**(3):739-744

[19] Buckel W. On the road to bioremediation of 'dioxin'. Chemistry & Biology. 2005;**12**:723-724

[20] Mitchell WJ. Physiology of carbohydrate to solvent conversion by Clostridia. Advances in Microbial Physiology. 1997;**39**:31-130

[21] Johnson EA. Clostridial toxins as therapeutic agents: Benefits of nature's most toxic proteins. Annual Review of Microbiology. 1999;**53**:551-575

Section 3

Biodiesel Catalysts

Chapter 4

Recent Developments in Catalysts for Biodiesel Production Applications

Teketel Alemu and Anshebo Getachew Alemu

Abstract

The world's most urgent problem today is the quick depletion of energy resources, which necessitates research into alternative energy sources in order to meet the world's explosive growth in energy demand. Among other renewable energy sources, biodiesel holds promise for meeting energy demand at a low cost through a variety of processes. In the biodiesel industry, sophisticated catalysts have recently grown in popularity for their ability to activate esterification and transesterification processes. The goal of this chapter is to give a general overview of catalyst developments, including their benefits and drawbacks in the biodiesel production process. In particular, we present a comparison of various homogeneous and heterogeneous catalysts. We found that nanocatalysts hold the most promise for the production of biodiesel.

Keywords: biodiesel, homogenous, heterogeneous, nanocatalysts, biodiesel production

1. Introduction

The increased global energy demand and the pursuit of environmentally friendly technology drive researchers toward alternative energy sources [1]. Currently, crude oil (35%), coal (29%), natural gas (24%), nuclear energy (7%), and renewable energy (5%) account for the majority of global energy consumption from fossil fuels [2]. This increased use of fossil fuels contributes to the global collapse of fossil fuels, air pollution, and global warming. Furthermore, it is predicted that all fossil fuel sources will be depleted by 2050 [3]. As a result, scholars have been motivated to find out renewable energy sources as clean energy alternatives, such as solar, wind, tidal, and geothermal energy, as well as biomass derived energy, to overcome these energy limitation [4]. Numerous innovative ideas can improve renewable energy technologies and provide sustainable methods to meet rising energy demand in a clean environment. In this regard, biodiesel is a potential fuel with a lower use of fossil fuels [5].

Furthermore, biodiesel offers the same performance as engine stability as petroleum diesel fuel, is nonflammable and nontoxic, reduces tailpipe emissions, visible smoke, and noxious fumes, and is safe for use in all conventional diesel engines [6]. It is produced from mono-alkyl esters of long-chain fatty acids through transesterification of vegetable oil by using catalysts as shown in **Figure 1**. It is renewable, nontoxic, biodegradable, and environmentally friendly and can be used in

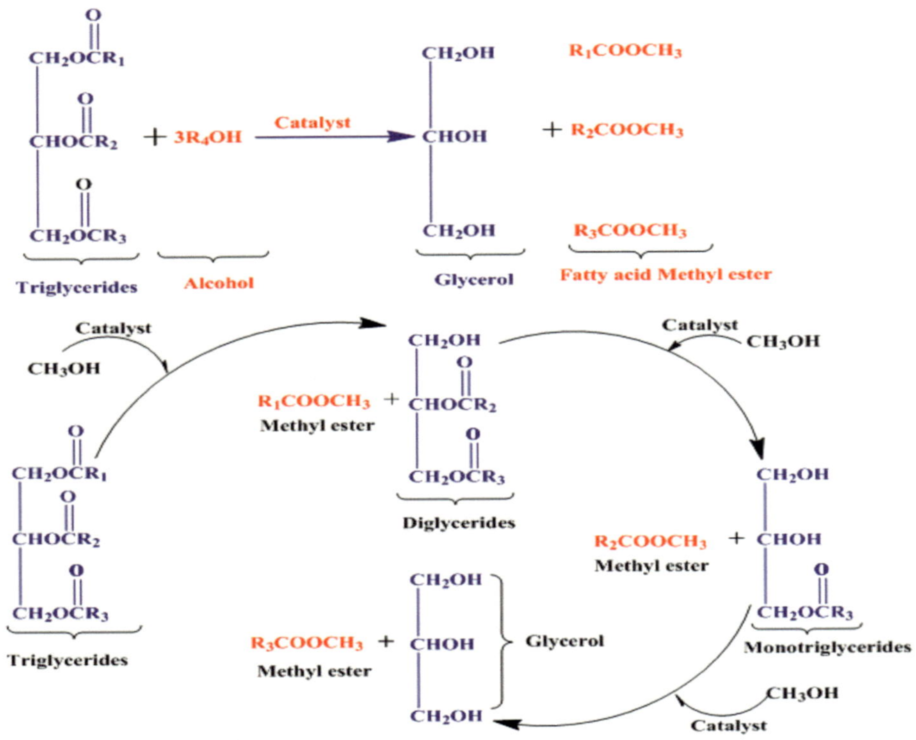

Figure 1.
Transesterification reactions of glycerides with alcohol to get methyl esters [7, 8].

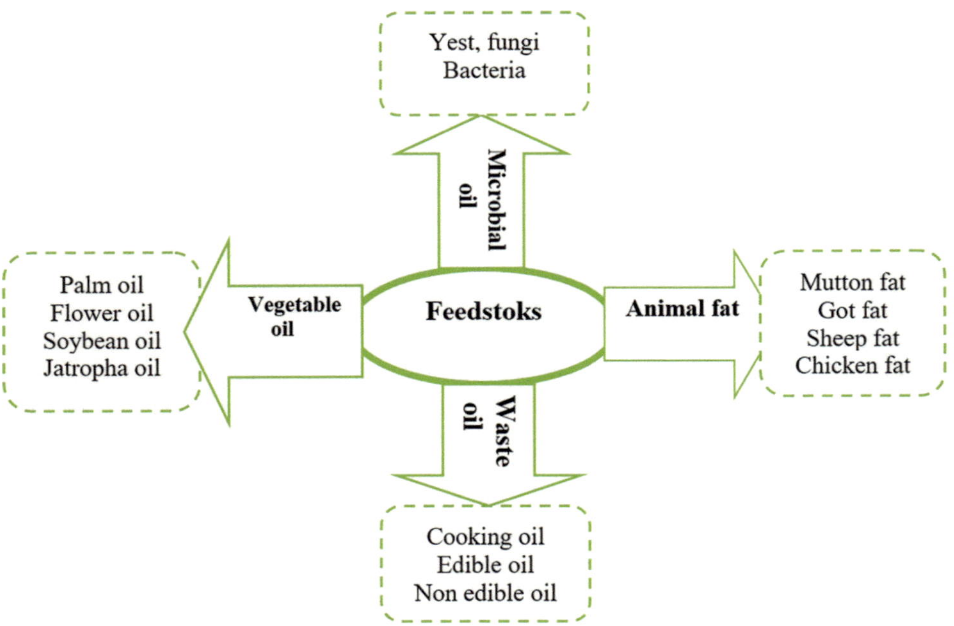

Figure 2.
Different feedstocks for biodiesel production.

compression-ignition (diesel) engines with little or no modification due to its adjustable physical and chemical properties. It is a renewable energy source made from various oil sources, such as animal fat, canola, mustard, soybean, and sunflower, waste oils, such as waste edible or nonedible oil, waste cooking oil, and microbial oil.

Figure 2 shows various sources of feedstock for biodiesel production. Furthermore, essential microorganisms in biodiesel production are classified into four types: bacteria, fungi, microalgae, and yeasts [7]. Yeasts include *Rhodotorula graminis* [8], *Candida tropicalis*, and *Yarrowia lipolytica* [9]. Fungi also include *Coniochaeta hoffmannii*, [10] *Alternaria alternata*, *Cladosporium cladosporioides*, *Epicoccum nigrum*, *Ffusarium oxysporum*, *Aspergillus parasiticus*, and *Emericella nidulans* [11]. *Chlorella minutissima* [12], *Scenedesmus obliquus*, and *Desmodesmus* spp. [13] are among the microalgae. Furthermore, bacteria such as *Serratia* sp. and *Bacillus* sp. were used [14–17]. Additionally, biodiesel is an ethyl ester derived from plant oil, microbial oil, animal oil, and the disposal of edible oil [16, 18].

2. Biodiesel production methods

Generally, there are many ways for biodiesel production. Recently, the most commonly used methods are esterification and transesterification. According to recent reports, esterification stands for the reaction of FFAs and alcohol to make FAAEs, and water is released, whereas transesterification stands for the reaction of triglycerides or triacylglycerols (TAGs) with alcohol to make FAAE and glycerol is produced as a by-product [16, 18]. In terms of reaction rate, esterification process is faster than transesterification process, for the reason that of it is short step reaction as shown in **Figure 1** [19, 20].

3. Catalysts for biodiesel production

Recently, catalysts have drawn a lot of attention for their applications in a range of fields, including materials science, bioremediation, food, photonic crystals, cosmetics, medication delivery, and food [19]. Nanomaterial-based catalysts have enormous promise for reducing biodiesel production costs and hastening transesterification processes [21]. The transesterification process is researched mostly through the employment of various types of nanocatalysts. The core, the surface, and the shell are the three layers that make up a nanoparticle, which is not a straightforward molecule. The NP's central region is known as the core, and its surface layer is made up of small molecules [8, 20, 22].

This chapter aims to overview recent developments in catalyst research and the types used in this process and their application to the production of biodiesel. Then, we point out the limitations of nanocatalysts in the production of biodiesel.

3.1 Biocatalysts (Enzyme)

Generally, either catalytic or non-catalytic processes are employed in the production of biodiesel. Among catalytic techniques, biocatalysts can be a successful way to make biodiesel in a sustainable way. Biocatalytic technologies for making biodiesel are in urgent demand to mitigate greenhouse gas emissions from conventional diesel or fossil fuels. The transesterification process in the production of biodiesel is usually catalyzed by lipases with superior biochemical and physiological features. In total,

70–95% of ethanol and methanol are produced by bacterial and fungal lipases [22]. Biodiesel is typically made of fatty acid alkyl esters, which are either mono-alkyl esters of fatty acid methyl esters or fatty acid ethyl esters, depending on the alcohol employed in the synthesis as shown in **Figures 1** and **3**.

Furthermore, the problems brought on by alkali and acid-catalyzed procedures are minimized by the biocatalytic biodiesel synthesis process. Since enzymes can esterify low-quality feedstock with a high concentration of free fatty acids, using enzyme catalysts has a number of economic and environmental advantages, such as the production of pure and highly valuable glycerol, the generation of the least amount of wastewater, the use of mild reaction conditions, and the absence of soap formation (FFA). Because they are simple to separate from the reaction mixture and have a generally lower chance of contamination, this production process is uncomplicated and has a low energy consumption [23, 24]. Recent developments in biomaterial catalysts, including cellulose, cellobiose, glucosidase, laccase, and xylanase, increase the effectiveness and durability of catalytic processes. Since nano biocatalysts may easily be reused and recovered by a continuous and large-scale process, using biocatalysts makes biocatalyst recovery and reusability easy (**Figure 4**) [24, 25].

3.2 Homogeneous catalysts

The most common biodiesel productions process, such as esterification, ester hydrolysis, and transesterification, have all been studied by using both acid and base homogeneous catalysts. A chemical in a like phase of the reaction structure catalyzes a series of reactions in production kinetics. The homogeneous catalyst is the most popular catalyst used because it is easy to react, good conversion rate, and very fast to complete the reaction in the synthesis of biodiesel. In order to dissolve homogeneous

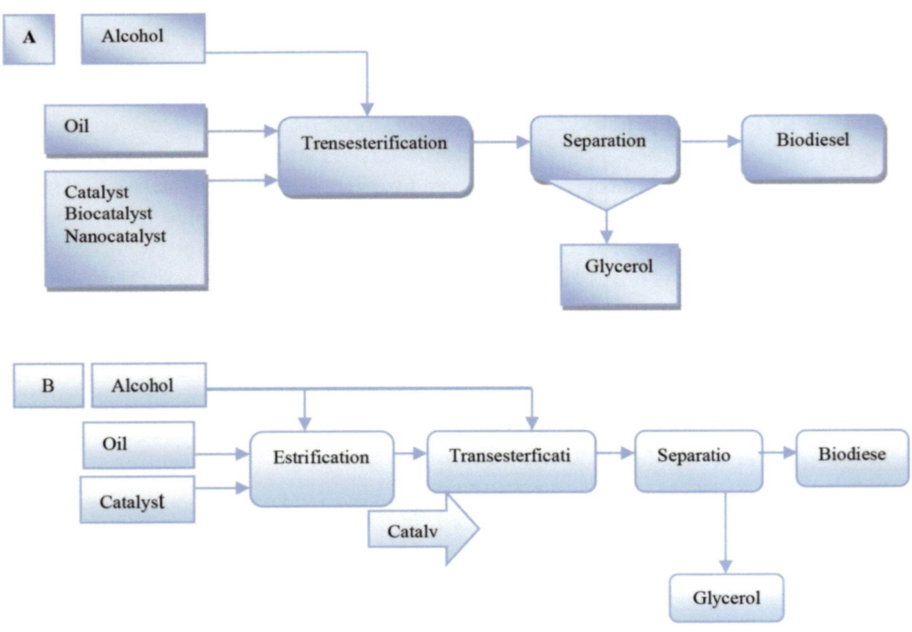

Figure 3.
Biodiesel production from feedstock: A. one-step process; B. two-step process [7, 8].

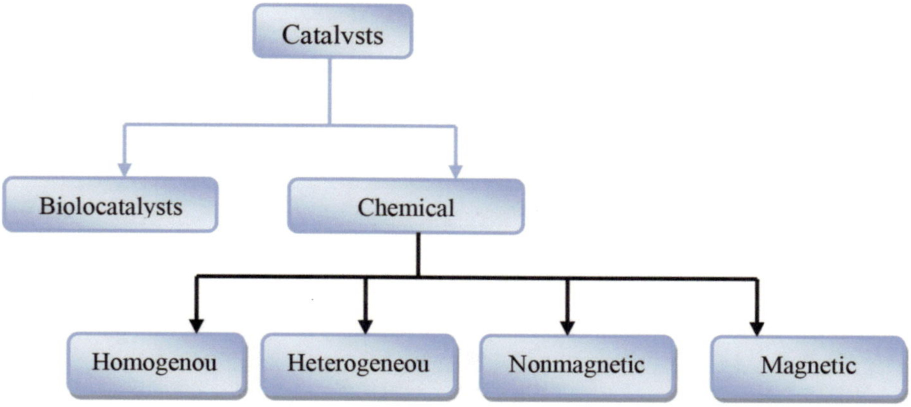

Figure 4.
The various types of catalysts.

catalysts, the solvent is normally in the same phase as all of the reactants. However, the fundamental limitation it is problematic for reuse, and sometimes recycling this catalyst is very expensive [25, 26].

3.2.1 Homogeneous alkaline (base) catalysts

These catalysts are mostly alkaline liquids with high activity in transesterification reaction of low free fatty acid (FFA) [25, 26]. The extensively used catalysts, such as sodium hydroxide (NaOH), potassium hydroxide (KOH), potassium methoxide (CH_3OK), and carbonates (**Table 1**) [27–43]. The conservative method for producing biodiesel from pure vegetable oils is the transesterification process using alkaline catalysts. For example, three moles of alcohol and one mole of triglyceride undergo this reaction, producing one mole of glycerol and three moles of fatty esters. The best alcohol-to-triglyceride ratio was a six-to-one give-up, which was 98%. Alkali metal methoxides are more energetic than alkali metal hydroxides, for the production of biodiesel, but its drawback is the significant amount of FFA remains as soap, it requests an extra catalyst and the catalyst is lost to the soap [44, 45]. In addition, deactivation, slow reaction rates, and saponification are the main drawback of alkaline catalysts, difficult to discern the product from biodiesel in the reaction system have greater than 2.5% FFA concentration result unwanted saponification, leads to a loss in enzymatic activity and requires extra energy to solve various technical issues [46, 47].

3.2.2 Homogeneous acid catalyst

The homogenous acid catalysts, such as hydrochloric acid (HCl), sulfonic($R\text{-}SO_3H$), and sulfuric acids (H_2SO4), as well as Brønsted acids are commonly used catalysts for both the esterification and transesterification process (**Table 1**). These catalysts increase the yield of alkyl esters and lesser the cost of the feedstock, making the process more cost-effective. However, the process is less appealing since it requires high temperatures, operates more slowly, causes corrosion, and has higher purifying and separation expenses. Less rate reactions are involved in the transesterification processes catalyzed by acids. However, due to the slower reaction rates compared to alkali-catalyzed reactions and the higher energy demands, the procedure is economically difficult [48, 49].

Catalyst	Feedstock	Yield/Conversion (%)	Ref
MgO	Moringa oleifera seeds oil	93.69	[27]
CaO-MgO	Jatropha curcas oil	90	[28]
CaO-ZnO	Sunflower oil	>.90	[29]
CaO-La_2O_3	Soybean oil	94.3	[30]
KF-CaO-Fe_3O_4	Stillingia oil	>95	[31]
KOH/AL_2O_3	Palm oil	91.07	[32]
Li/ZnO-Fe_3O_4	Rapeseed oil	99.8	[33]
KF/CaO	Tallow seed oil	96.8	[34]
$NaAlO_2$/g-Al_2O_3	Palm oil	97.65	[35]
25%MoO_3/B-ZSM-5	Oleic acid	98	[36]
CaO/CuFe2O4	Chicken fat	94.52	[37]
KOH/Fe_3O_4@Al_2O_3	Canola oil	98.8	[38]
MgO/$MgFe_2O_4$	Sunflower oil	91.2	[39]
Cr/Ca/g-Al_2O_3	Cooking oil	92.79	[40]
γ-Al2O_3/KI	Palm oil	98	[41]
Ca/γ-Al2O_3	Corn oil	87.89	[42]
Cs/Al/Fe3O_4	Sunflower oil	95	[43]

Table 1.
Various nanocatalysts are used for biodiesel production from different sources [27–43].

Therefore, homogeneous acid catalysis is applicable for both transesterification and esterification reaction processes [50, 51]. For example, p-toluene sulphonic acid catalyst and the Amberlyst-35 sulphonic acid catalyst were employed in series at 5% concentration in a specific application to extract the ester from vegetable refining waste. There are various alcohols, such ethanol, butanol, and methanol, for esterification reactions producing >90% biodiesel [52].

3.3 Heterogeneous catalysts

These groups of catalysts are involved in a different state of product and reactants and also, they are noncorrosive, as well as easily separable from the products. These catalysts offer certain exceptional qualities. Furthermore, heterogeneous catalysts (solid) have empathetic nature with little harm to the environment. It is crucial to develop some suitable heterogeneous catalysts for the manufacture of biodiesel from affordable feedstocks. These heterogeneous catalysts are further subdivided into basic and acidic catalysts [51, 52].

3.3.1 Heterogeneous base catalysts

Heterogeneous base catalysts overcome a number of obstacles, including saponification, which prevents glycerol from separating from the methyl ester layer. Heterogeneous base catalysts overcome a number of obstacles, including saponification, which prevents glycerol from separating from the methyl ester surface. In contrast, these catalysts have promising advantages such as environmental friendliness, reduced waste material harm, non-corrosiveness, selectivity, tolerance of high

FFA and moisture contents, promoting simple recovery, reusability, low cost, and green process. They can also be modified to increase activity, selectivity, and catalyst lifetime [53, 54]. Base alkali earth metal oxides, such as BeO, MgO, CaO, SrO, BaO, and RaO [55, 56], transition metal oxides [57], mixed metal oxides, such as $CaTiO_3$, $CaMnO_3$, Ca_2Fe2O_5, $CaZrO_3$, and CaO-CeO_2, ion exchange resins, and alkali metal compounds based on alumina are the most useful [58–61]. These solid-base catalysts, including CaO, MgO, SrO, KNO_3/Al_2O_3, K_2CO_3/Al_2O_3, KF/Al_2O_3, Li/CaO, and KF/ZnO, applicable for transesterification [62, 63]. The basic hydrotalcite of Mg/Al, Li/Al, anion exchange resins, base zeolites, hydrotalcite, calcium carbonate rock, Li/CaO, MgO/KOH, and Na/NaOH/-Al_2O_3 [64–67]. The oxide catalysts exhibit high yields and stability in the transesterification process [68, 69].

3.3.2 Heterogeneous acid catalysts

These catalysts are less toxic, corrosive, and cause less environmental issues [70]. They have a variety of acidic sites with varying levels of Lewis acidity. Despite offering encouraging results under modest reaction circumstances, they react much more slowly than solid base catalysts. These kinds of catalysts have further conditions like a high catalyst loading, high temperature, and prolonged reaction time [71]. Additionally, solid acid catalysts support the simultaneous transesterification and esterification of oils with high FFA contents to produce biodiesel. For example, solid-acid catalysts with organo-sulfonic groups, such as Nafion and Amberlyst, are used to speed up the esterification of fatty acids. In the transesterification reaction, another mesostructured catalyst modified by sulphonic acid is employed, leading to conversion rates as high as 100% [56, 72–74].

3.3.3 Heterogeneous nanocatalysts

These catalysts are known to improve the rate of transesterification reaction by removing unwanted processes and unnecessary reaction yield. These catalysts advance simple recovery, reusability, and a cost-effective friendly process [75]. In addition, these catalysts exhibit a number of advantages, such as enduring high FFA and moisture content, essential in certain insensitive high temperature and pressure. The cost-effective heterogeneous catalysts help out to diminish the overall cost of biodiesel production. These catalysts may be made to have a maximum yield of a reaction product by altering the number of atoms, surface functionality, and elemental composition, and they also have an efficient surface area, high stability, and higher resistance to saponification [76–80]. There are variety of techniques, including vacuum deposition, self-propagating high-temperature synthesis, evaporation, coprecipitation, electrochemical deposition, microwave combustion, hydrothermal, solvothermal, impregnation, and sol–gel technology [56, 81]. These catalysts are formed from nanoparticles with less than 100 nm variety of sizes and morphologies. They demonstrate critical advantages for both heterogeneous and homogeneous catalysts in terms of activity, selectivity, efficiency, and reusability [82, 83]. Here, biodiesel nanocatalysts are divided into magnetic and nonmagnetic categories.

3.3.3.1 Magnetic nanocatalysts

These catalysts can aid in the reaction without the need for centrifugation or ultra-filtration. It is a powerful tool for the rapid separation of catalysts from reaction systems, offering an alternative to time-consuming, solvent-intensive, and energy-intensive separation procedures while sustaining catalytic activity for successive cycles. They are suitable for inexpensive feedstocks [84, 85]. Among the number of magnetic catalyst Fe_3O_4, Cao/Fe_3O_4,$Ca(OH)_2$/Fe_3O_4,Cs/Al/Fe_3O_4,KF/CaO-Fe_3O_4,Fe_3O_4@SiO_2,MgO/$MgFe_2O_4$, and Ca/Fe_3O_4@SiO_2 are few of the magnetic catalyst that have recently been developed and used to make biodiesel [86–89]. For example, A. Ali et al. reported the use of the CaO-Fe_3O_4 magnetic catalyst in the generation of biodiesel from palm seed oil. In addition, cadmium oxide and tin oxide magnetic nanocatalysts have been employed for esterification, transesterification, and hydrolysis reactions of soybean oil [90]. Furthermore, more active catalysts Fe_3O_4 and Fe_3O_4@SiO_2 (MNPs) have been used as fundamental recyclable catalysts achieving 96% yield production. The catalyst ZnO/$BiFeO_3$ was also a promising catalyst for the generation of biodiesel from canola oil and yields of 95.43 and 95.02% in the first and second cycles, respectively [90, 91]. Another example, SnO produces 84% yield of esterification without loss, at 200°C after 1 h reaction. Therefore, magnetic catalyst show fast, clean, superior stability, and recyclability.

3.3.3.2 Nonmagnetic nanocatalysts

The ability to reuse the catalysts allows for lower-cost biodiesel production in a fixed-bed reactor. There are number of nonmagnetic nanocatalysts, such as hydrotalcite, metal oxides, sulfated oxides, zeolites, and zirconia, which are frequently employed in the manufacture of biodiesel [44, 77, 92–95]. Molina also used ZnO nanorods to produce biodiesel from olive oil and realized that their catalytic performance yield of 94.8% was slightly higher than the regular ZnO yield of 91.4% [81, 96]. Borah et al. reported a maximum production of 98.03% at a methanol/oil ratio of 9 at 60°C for 3 hours with 2.5 wt% Co/ZnO catalysts for biodiesel [34]. Zhang et al. improved by using the surface modification of the $NaAlO_2$/c-Al_2O_3 with the M/O of 20.79:1, 10.89 wt% catalysts, and at 64.72°C to achieve the highest biodiesel production, which was 97.65% [97]. Alkaline earth metal compounds, in particular, Ca-enclose nanomaterials, have the potential to be nonmagnetic catalysts for the transesterification of biodiesel. The most widely used nonmagnetic catalysts include MgO/TiO_2 [98], Mg-Al hydrotalcite [99], KF/CaO [100], Mg/Al [101], Li/CaO [102], MgO [40], metal–organic frameworks (MOFs) [97, 103], Nanozeolites al. [104], potassium-doped zeolite [27], and hydrotalcite. The implementation of nanocatalysts gets much attention due to vast surface area strong catalytic performance, and appropriate charge transport channel.

4. Advantages and disadvantages of different catalysts

Finally, **Table 2** highlights the advantages and disadvantages of various catalysts. From this point of view, a homogeneous catalyst has been thoroughly examined, and drawbacks have been discussed in the literature. However, there is currently a lot of study being done in the relatively new field of heterogeneous catalysts. The literature has noted a number of benefits and drawbacks for catalysts, as summarized in **Table 2**.

Types of catalyst	Advantages	Disadvantages	Ref.
Homogenous base catalysts	• Simple to use, require less time to achieve a complete reaction high selectivity, turnover frequency, reaction effortless optimization of activity, dissolves quickly.	• Containing significant amounts of FFA, cannot be converted into biodiesel completely, remains as soap in vast quantities, required additional amount of catalyst and catalyst lost, sensitivity toward moisture and free fatty acids.	[105–107]
Homogenous acid catalysts	• Offers very high yield, optimum reaction conditions, the process economically confronting, the increased energy requirements.	• Slow process, it requires high temperature, it causes corrosion (reactors, pipes, vessels, etc.), other environmental issues, product separation, and purification costs make it a less-attractive process.	[49–52]
Heterogeneous acid catalysts	• Ease of separation. • Catalysts can be reused.	• High cost of synthesizing catalyst. • Higher alcohol-to-oil molar ratios. • High catalyst concentrations. • Longer reaction time. • May undergo deactivation.	[44–46]
Heterogeneous base catalysts	• Ease of purification of the product. • Effluent generation minimized. • Catalysts can be reused.	• High cost of synthesizing catalyst. • Leaching of active sites may occur.	[53–55, 105–107]
Magnetic nanocatalysts	• More active. • Suitable for biodiesel production from low-cost feedstocks. • Reused without loss in the activity. • Cost and energy savings, particularly for high-temperature catalytic reactions. • Maintain stability after mechanical, physical, and chemical modification.	• Poor dispersion abilities. • High cost of synthesis material • Limitations in scale-up production processes. • Mobility dependent on environment compatibilities. • Limited availability for the catalytic process.	[56, 81–87]
Nonmagnetic nanocatalysts	• Lower-cost production. • Operate in a fixed-bed reactor. • More suitable for the process.	• Takes a long time to proceed with the reaction.	[88–93, 103]
Biocatalyst	• Ease of separation. • Mild reaction condition. • Produces high-purity product. • Insensitive to FFA and water content. • In the oil	• High cost for catalyst synthesis. • Sensitive to methanol, causing • deactivation of the enzyme. • Very slow reaction rate.	[23, 24]

Types of catalyst	Advantages	Disadvantages	Ref.
Nanocatalyst	• High activity and stability. • High reusability. • Mild reaction condition.	• High cost for catalyst synthesis.	[2, 56, 73–76, 79, 80]

Table 2.
Different group of catalysts for biodiesel production.

5. Conclusion

The development of various catalysts has recently received a lot of attention due to their high operating catalytic effectiveness. Numerous catalysts have been studied for their ability to improve biodiesel production performance. Among the various catalysts, homogeneous and heterogeneous are the better choices for biodiesel production, and their reaction is primarily dependent on catalytic systems. As a result, heterogeneous catalysts surpass homogeneous catalysts in terms of ease of separation, simplicity, and reusability. To overcome limitations such as stability, recyclability, durability, and aggregation, the production of highly efficient catalysts necessitates additional innovative strategies. The development of extremely active and selective catalysts, as well as their economic viability for industrial use, is an issue that must be addressed. The development of extremely active and selective catalysts that are also economically viable for use in biodiesel production is priority must be addressed.

Acknowledgements

The authors acknowledge Intech open access for the invitation and constructive comments.

Conflict of interest

The authors declare no conflict of interest.

DOI: http://dx.doi.org/10.5772/intechopen.109483

Author details

Teketel Alemu[1] and Anshebo Getachew Alemu[2]*

1 College of Engineering and Technology, Wachemo University, Department of Chemical Engineering, Ethiopia

2 Facultyof Natural and Computational Science, Department of Physics, Samara University, Ethiopia

*Address all correspondence to: agetachew2013alemu@yahoo.com

References

[1] Sharma V et al. Transesterification of Pyrolysed Castor Seed Oil in the Presence of CaCu(OCH3) 2 Catalyst. Energies. 2021;**14**:6064

[2] Thangaraj B et al. Catalysis in biodiesel production—A review. Clean Energy. 2019;**3**(1):2-23

[3] Esmaeili H. Application of nanotechnology for biofuel production. Emerging Nanotechnologies for Renewable Energy. 2021;**149**:172

[4] Hussein AK. Applications of nanotechnology, in renewable energies—A comprehensive overview and understanding. Renewable and Sustainable, Energy Reviews. 2015;**42**:460

[5] Bano S et al. Fabrication and optimization of nanocatalyst for biodiesel production: An overview. Frontiers in Energy Research. 2020;**8**:579014

[6] Chand N. Plant oils - Fuel of the future. Journal of Scientific and Industrial Research. 2002;**61**:7-16

[7] Ingle AP et al. Advances in nanocatalysts mediated biodiesel production: A critical appraisal. Symmetry. 2020;**12**:256

[8] Martinez-Silveira A et al. Process conditions for a rapid in situ transesterification for biodiesel production from oleaginous yeasts. Electronic Journal of Biotechnology. 2019;**38**:1

[9] Louhasakul Y et al. Potential use of flocculating oleaginous yeasts for bioconversion of industrial wastesinto biodiesel feedstocks. Renewable Energy. 2019;**136**:1311

[10] Karatay SE, Demiray E, Dönmez G. Efficient approaches to convert Coniochaet ah off Manniilipids into biodieselby in-situ transesterification, Bioresource Technology. 2019;**285**:121321

[11] Bagy MMK et al. Two stage biodiesel and hydrogen production from molasses by oleaginous fungi and Clostridium acetobutylicum ATCC. International Journal of Hydrogen Energy. 2014;**39**:3185

[12] Pereira FM et al. Evaluation of fatty acids production by Chlorella minutissima in batch bubble- column photobioreactor. Fuel. 2018;**230**:155

[13] Hernández-García A et al. Wastewater-leachate treatment by microalgae: Biomass, carbohydrate and lipid production. Ecotoxicology and Environmental Safety. 2019;**174**:435-444

[14] Kumar M, Thakur IS. Municipal secondary sludge as carbon source for production and characterization of biodiesel from oleaginous bacteria. Bioresource Technology Reports. 2018;**4**:106-113

[15] Cea M et al. Screening transesterifiable lipid accumulating bacteria from sewage sludge for biodiesel production. Biotechnology Reports. 2015;**8**:116-123

[16] Moser BR. Biodiese production, properties, and feedstocks. In Vitro Cellular & Developmental Biology - Plant. 2009;**45**:229-266

[17] Fattah IMR, Masjuki HH, Kalam MA, Wakil MA, Rashedul HK, Abedin MJ. Performance and emission characteristics of a CI engine fueled with Cocos nucifera and Jatropha curcas B20 blends

accompanying antioxidants. Industrial Crops and Products. 2014d;**57**:132

[18] Pasupulety N et al. Production of biodiesel from soybean oil on CaO/Al2O3 solid base catalysts. Applied Catalysis A: General. 2013;**452**:189-202

[19] Lee YC, Lee K, Oh YK. Recent nanoparticle engineering advances in microalgal cultivation and harvesting processes of biodiesel production: A review. Bioresource Technology. 2015;**184**:63-72

[20] Khan I, Saeed K, Khan I. Nanoparticles: Properties, applications and toxicities. Arabian Journal of Chemistry. 2019;**12**(7):908-931

[21] Laurent D et al. Magnetic iron oxide nanoparticles: Synthesis, stabilization, vectorization, physicochemical characterizations, and biological applications. Chemical Reviews. 2008;**108**(6):2064

[22] Zik NAFA, Sulaiman S, Jamal P. Thermally produced nano catalyst for biodiesel production: A review. Journal of Advanced Research in Fluid Mechanics and Thermal Sciences. 2018;**52**:139-147

[23] Jin G, Bierma TJ, Hamaker CG, Rhykerd R, Loftus LA. Producing biodiesel using whole-cell biocatalysts in separate hydrolysis and methanolysis reactions. Journal of Environmental Science and Health, Part A Toxic/Hazardous Substances and Environmental Engineering. 2008;**43**:589-595

[24] Akoh CC, Chang S, Lee G, Shaw J. Enzymatic approach to biodiesel production. Journal of Agricultural and Food Chemistry. 2007;**55**:8995

[25] Banković Ilić IB, Miladinović MR, Stamenković OS, Veljković VB. Application of nano CaO–based catalysts in biodiesel synthesis. Renewable and Sustainable Energy Reviews. 2017;**72**:746

[26] Salimi Z, Hosseini SA. Study and optimization of conditions of biodiesel production from edibleoils using ZnO/BiFeO3 nano magnetic catalyst. Fuel. 2019;**239**:1204

[27] Esmaeili H, Yeganeh G, Esmaeilzadeh F. Optimization of biodiesel production from Moringaoleifera seedsoil in the presence of nano-MgO using Taguchi method. International Nano Letters. 2019;**9**:257

[28] Taufiq YY. Transesterification of non-edible Jatropha curcas oil to biodiesel using binary Ca-Mg mixed oxide catalyst: Effect of stoichiometric composition. Chemical Engineering Journal. 2011;**178**:342

[29] Rubio CJM. Calcium zincate as precursor of active catalysts for biodiesel production under mild conditions. Applied Catalysis B: Environmental. 2009;**91**:339-346

[30] Yan S. Oil transesterification over calcium oxides modified with lanthanum. Applied Catalysis A: General. 2009;**360**:163

[31] Kesi Z. Mechanochemical preparation and characterization of CaOZnO used as catalyst for biodiesel synthesis. Applied Catalysis A: General. 2012;**427**:58

[32] Hu S. Nano-magnetic catalyst KF/CaO_Fe3O4for biodiesel production. Applied Energy. 2011;**88**:2685

[33] Mithileth P, Shiao SC, Bhanupriya D, Hau MC, Chang C-T, Thi XQN, et al. Ultrasonication assisted catalytic transesterification of ceiba pentandra (kapok) oil derived biodiesel using immobilized iron nanoparticles. Fuels. 2022;**3**:113

[34] Wen L. Preparation of KF/CaO nanocatalyst and its application in biodiesel production from Chinese tallow seed oil. Fuel. 2010;**89**:2267-2271

[35] Zhang Y, Niu S, Lu C, Gong Z, Hu X. Catalytic performance of NaAlO2/g-Al2O3 as heterogeneous nanocatalyst for biodiesel production: Optimization using response surface methodology. Energy Conversion and Management. 2020;**203**:112263

[36] Mohebbi S, Rostamizadeh M, Kahforoushan D. Effect of molybdenum promoter on performance of high silica MoO_3/BZSM$_{-5}$ nanocatalyst in biodiesel production. Fuel. 2020;**266**:117063

[37] Seffati K, Honarvar B, Esmaeili H, Esfandiari N. Enhanced biodiesel production from chicken fat using CaO/CuFe2O4 nanocatalyst and its combination with diesel to improve fuel properties. Fuel. 2019;**235**:1238

[38] Kazemifard S, Nayebzadeh H, Saghatoleslami N, Safakish E. Assessment the activity of magnetic KOH/Fe3O4@ Al2O3 core–shell nanocatalyst in transesterification reaction: Effect of Fe/ Al ratio on structural and performance. Environmental Science and Pollution Research. 2018;**25**:32811

[39] Alaei S, Haghighi M, Toghiani J, Rahmani Vahid B. Magnetic and reusable MgO/$MgFe_2O_4$ nanocatalyst for biodiesel production from sunflower oil: Influence of fuel ratio in combustion synthesis on catalytic properties and performance. Industrial Crops and Products. 2018;**117**:322

[40] Sulaiman NF, Wan Abu Bakar WA, Ali R. Response surface methodology for the optimum production of biodiesel over Cr/Ca/g-Al2O3 catalyst: Catalytic performance and physicochemical studies. Renewable Energy. 2017;**113**:697-705

[41] Islam A, Taufiq Yap YH, Ravindra P, Teo SH, Sivasangar S, Chan ES. Biodiesel synthesis overmillimetric g-Al2O3/KI catalyst. Energy. 2015;**89**:965

[42] Moradi G, Mohadesi M, Rezaei R, Moradi R. Biodiesel production using CaO/g-Al2O3 catalyst synthesized by sol-gel method. The Canadian Journal of Chemical Engineering. 2015;**2015**(93):1531

[43] Feyzi M, Hassankhani A, Rafiee HR. Preparation and characterization of Cs/ Al/Fe3O4 nanocatalysts for biodiesel production. Energy Conversion and Management. 2013;**71**:62

[44] Zhang Q, Ling D, Lei D, Wang J, Liu X, Zhang Y, et al. Green and facile synthesis of metal-organic framework Cu-BTC-supported Sn (II)-substituted Keggin Heteropoly composites as an esterification nanocatalyst for biodiesel production. Frontiers in Chemistry. 2020;**8**:129

[45] Zhang D, Wei D, Li Q, Ge X, Guo X, Xie Z, et al. High performance catalytic distillation using CNTs-based holistic catalyst for production of high quality biodiesel. Scientific Reports. 2014;**4**:4021

[46] Mouaimine M, Andriano JD, Fidele PT, Pierre V, Mohamed MS. Plant latex lipase as biocatalysts for biodiesel production. African Journal of Biotechnology. 2016;**15**:1487

[47] Zhang QY, Ling D, Lei DD, Deng TL, Zhang YT, Ma PH. Synthesis and catalytic properties of nickel salts of Keggin type heteropoly acids embedded metal-organic framework hybridnanocatalyst. Green Processing and Synthesis. 2020;**9**:131

[48] Silitonga AS, Shamsuddin AH, Mahlia TMI, Milano J, Kusumo F, Siswantoro J, et al. Biodiesel synthesis

from Ceiba pentandra oil by microwave irradiation-assisted transesterification: ELM modeling and optimization. Renewable Energy. 2020;**146**:1278

[49] Fukuda H, Kondo A, Noda H. Biodiesel fuel production by transesterification of oils. Journal of Bioscience and Bioengineering. 2001;**92**:405

[50] Zheng S, Kates M, Dube MA, et al. Acid-catalyzed production of biodiesel from waste frying oil. Biomass and Bioenergy. 2006;**30**:267

[51] Marchetti JM, Miguel VU, Errazu AF. Possible methods for biodiesel production. Renewable and Sustainable Energy Reviews. 2007;**11**:1300

[52] Reis MC, Freitas FA, Lachter ER. Biodiesel production from fatty acids of refined vegetable oils by heterogeneous acid catalysis and microwave irradiation. Química Nova. 2015;**38**:130

[53] Calero J, Luna D, Sancho ED, Luna C, Bautista FM, Romero AA. Development of a new biodiesel that integrates glycerol, by using CaO as heterogeneous catalyst, in the partial Methanolysis of sunflower oil. Fuel. 2014;**122**:94

[54] Liu X, He H, Wang Y, Zhu S, Piao X. Transesterification of soybean oil to biodiesel using CaO as a solid base catalyst. Fuel. 2008b;**87**:216

[55] Refaat AA. Different techniques for the production of biodiesel from waste vegetable oil. International Journal of Environmental Science & Technology. 2010;7:183

[56] Liu R, Wang X, Zhao X. Sulfonated ordered mesoporous carbon for catalytic preparation of biodiesel. Carbon. 2008;**46**:1664-1669

[57] Corma A, Iborra S. Optimization of alkaline earth metal oxide and hydroxide catalysts for base-catalyzed reactions. Advances in Catalysis. 2006;**49**:239-302

[58] Chouhan APS, Sarma AK. Modern heterogeneous catalysts for biodiesel production: A comprehensive review. Renewable and Sustainable Energy Reviews. 2011;**15**:4378

[59] Jørgensen KA. Transition-metal-catalyzed epoxidations. Chemical Reviews. 1989;**89**:431

[60] Gawande MB, Pandey RK, Jayaram RV. Role of mixed metaloxides in catalysis science— Versatile applications in organicsynthesis. Catalysis Science & Technology. 2012;**2**:1113

[61] Dantas J, Leal E, Cornejo DR, et al. Biodiesel production evaluating the use and reuse of magnetic nanocatalysts Ni0.5Zn0.5Fe2O4. Arabian Journal of Chemistry. 2018. DOI: 10.1016/j.arabjc.2018.08.012

[62] Dos Reis SCM, Lachter ER, Nascimento RSV, et al. Transesterification of Brazilian vegetable oils with Methanol over ion-exchange resins. Journal of the American Chemical Society. 2005;**82**:661

[63] Harmer MA, Sun Q. Solid acid catalysis using ion-exchange resins. Applied Catalysis A: General. 2001;**221**:45

[64] Gama PE, Rass G, Lachter ER. Biodiesel production by in situ transesterification of sun flower seeds by homogeneous and heterogeneous catalysis. Quim Nova. 2010;**1859**:33

[65] Gama PE, Lachter ER, Gil RASS. Characterization and catalytic activity of K_2CO_3/Al_2O_3 in the transesterification of sunflower oil in conventional and microwave heating. Quim Nova. 2015;**38**:185

[66] Wan T, Yu P, Wang S, et al. Application of sodium aluminate as a heterogeneous base catalyst for biodiesel production from soybean oil. Energy & Fuels. 2009;**23**:1089

[67] Vyas AP, Verma JL, Subrahmanyan N. A review on FAME production processes. Fuel. 2010;**89**:1

[68] Xie W, Li H. Alumina-supported potassium iodide as a heterogeneous catalyst for biodiesel production from soybean oil. Journal of Molecular Catalysis A: Chemical. 2006;**255**:1

[69] İlgen O, Akin AN. Development of alumina supported alkaline catalysts used for biodiesel production. Turkish Journal of Chemistry. 2009;**33**:281

[70] Endalew AK, Kiros Y, Zanzi R. Inorganic heterogeneous catalysts for biodiesel production from vegetable oils. Biomass and Bioenergy. 2011;**35**:3787

[71] Kawashima A, Matsubara K, Honda K. Development of heterogeneous base catalysts for biodiesel production. Bioresource Technology. 2008;**99**:3439

[72] Aransiola EF, Ojumu TV, Oyekola OO, Madzimbamuto TF, Ikhu-Omoregbe DIO. A review of current technology for biodiesel production: State of the art. Biomass and Bioenergy. 2014;**61**:276

[73] Mansir N, Taufiq-Yap YH, Rashid U, Lokman IM. Investigation of heterogeneous solid acid catalyst performance on low grade feedstocks for biodiesel production: A review. Energy Conversion and Management. 2017;**141**:171

[74] Mbaraka IK, Shanks BH. Design of multifunctionalized mesoporous silicas for esterification of fatty acid. Journal of Catalysis. 2005;**229**:365

[75] Mbaraka IK, Shanks BH. Conversion of oils and fats using advanced mesoporous heterogeneous catalysts. Journal of the American Oil Chemists' Society. 2006;**83**:79

[76] Ambat I, Srivastava V, Sillanpää M. Recentadvancementinbiodieselproduction methodologies using various feedstock: A review. Renewable and Sustainable Energy Reviews. 2018;**90**:356

[77] Akia M, Yazdani F, Motaee E, Han D, Arandiyan H. A review on conversion of biomass to biofuel by nanocatalysts. Biofuel Research Journal. 2014;**1**:16-25

[78] Borah MJ, Devi A, Saikia RA, et al. Biodiesel production from waste cooking oil catalyzed by in-situ decorated TiO2 on reduced graphene oxide nanocomposite. Energy. 2018;**158**:881-889

[79] Polshettiwar V, Varma RS. Green chemistry by nanocatalysis. Green Chemistry. 2010;**12**:743

[80] Diercks CS, Liu Y, Cordova KE, Yaghi OM. The role of reticular chemistry in the design of CO2 reduction catalysts. Nature Materials. 2018;**17**:301

[81] Mguni LL, Meijboom R, Jalama K. Biodiesel production over nano-MgO supported on titania. International Journal of Chemical, Molecular, Nuclear, Materials and Metallurgical Engineering. 2012;**64**:380

[82] Ali MA, Al-Hydary IA, Al-Hattab TA. Nano-magnetic catalyst CaO-Fe3O4 for biodiesel production from date palm seed oil. Bulletin of Chemical Reaction Engineering and Catalysis. 2017;**12**:460

[83] Feyzi M, Hassankhani A, Rafiee H. Preparation and characterization of CsAlFe3O4 nanocatalysts for biodiesel production. Energy Conversion and Management. 2013;**71**:62-68

[84] Liu Ch, Lv P, Yuan Zh, Yan F, Luo W. The nanometer magnetic solid base

catalyst for production of biodiesel. Renewable Energy. 2010;**5**:1531

[85] Feyzi M, Norouzi L. Preparation and kinetic study of magnetic Ca/ Fe3O4@SiO2 nanocatalysts for biodiesel production. Renewable Energy. 2016;**94**:579

[86] Alves MB, Medeiros FCM, Sousa MH, Rubim JC, Suarez PAZ. Camium and tin magnetic nanocatalysts useful for biodiesel production. Journal of the Brazilian Chemical Society. 2014;**12**:2304

[87] Salimi Z, Hosseini SA. Study and optimization of conditions of biodiesel production from edible oils using ZnO/ BiFeO3 nanomagnetic catalyst. Fuel. 2019;**239**:1204

[88] Amini G, Najafpour GD, Rabiee SM, Ghoreyshi AA. Synthesis and characterization of amorphous nano–alumina powders with high surface area for biodiesel production. Chemical Engineering and Technology. 2013;**36**:1708

[89] Han H, Guan Y. Synthesis of biodiesel from rapeseed oil using K2O/c–Al2O3 as nano–solid–base catalyst. Wuhan University. 2009;**14**:75

[90] Boz N, Degirmenbasi N, Kalyon DM. Conversion of biomass to fuel: Transesterification of vegetable oil to biodiesel using KF loaded nano–c–Al2O3 as catalyst. Applied Catalysis B: Environmental. 2009;**89**:590

[91] Feyzi M, Shahbazia E. Catalytic performance and characterization of Cs–Ca/SiO2– TiO2 nanocatalysts for biodiesel production. Journal of Molecular Catalysis A: Chemical. 2015;**404**:131

[92] Faria EA, Marques JS, Dias IM, Andrade RDA, Suareza PAZ, Prado AG. Nanosized and reusable SiO2/ZrO2 catalyst for highly efficient biodiesel production by soybean transesterification. Journal of the Brazilian Chemical Society. 2009;**2**:1732

[93] Tamilmagan MA, Bijesh P, Gopal A. Biodiesel production from waste cooking oil using green synthesized nano Fe2O3 and CuO impregnated nano Fe3O4. International Journal of ChemTech Research. 2015;**8**:90

[94] Molina CMM. ZnO nanorods as catalyts for biodiesel production from olive oil. Electronic Theses and Dissertations. Paper 1000. 2013;**5**:1829

[95] Borah MJ, Devi A, Borah R, Deka D. Synthesis and application of Co doped ZnO as heterogeneous nanocatalyst for biodiesel production from non-edible oil. Renewable Energy. 2019;**133**:512

[96] Obadiah A, Kannan R, Ravichandran P, Ramasubbu A, Vasanth Kumar S. Nano hydrotalcite as a novel catalyst for biodieselconversion. Digest Journal of Nanomaterials and Biostructures. 2012;**7**:321

[97] Deng X, Fang Z, Liu YH, Liu YC. Production of biodiesel from Jatropha oil catalyzed by nanosized Solid Basic catalyst. Energy. 2011;**36**:777

[98] Kaur M, Ali A. Lithium ion impregnated calcium oxide as nano catalyst for the biodiesel production from karanja and jatropha oils. Renewable Energy. 2011;**36**:2866

[99] Wang L, Yang J. Transesterification of soybean oil with nano- MgO or not in supercritical and subcritical methanol. Fuel. 2007;**86**:328

[100] Zhang QY, Yang TT, Liu XF, Yue CY, Liufang A, Deng TL, et al. Heteropoly acid- encapsulated metal-organic framework as a stable and highly efficient nanocatalyst for esterification reaction. RSC Advances. 2019;**9**:16357

[101] Vasconcellos A, Miller AH, Aranda DAG, Nery JG. Biocatalysts based on nano zeolite-enzyme complexes: Effects of alkoxysilane surface functionalization and biofuel production using microalgae lipids feedstock. Colloids and Surfaces. B, Biointerfaces. 2018;**165**:150

[102] Saeedi M, Fazaeli R, Aliyan H. Nanostructured sodium– zeolite imidazolate framework (ZIF-8) doped with potassium by sol-gel processing for biodiesel production from soybean oil. Journal of Sol-Gel Science and Technology. 2016;**77**:404

[103] Lamberti M, Zappavigna S, Sannolo N, Porto S, Caraglia M. Advantages and risks of nanotechnologies in cancer patients and occupationally exposed workers. Expert Opinion on Drug Delivery. 2014;**11**:1087

[104] Rahmani VB, Haghighi M, Alaei S, Toghiani J. Reusability enhancement of combustion synthesized MgO/MgAl2O4 nanocatalyst in biodiesel production by glow discharge plasma treatment. Energy Conversion and Management. 2017;**143**:23

[105] Vivek P, Rafael L. Magnetically recoverable nanocatalysts. Chemical Reviews. 2011;**111**:3036

[106] Helwani Z, Othman MR, Aziz N, Kim J, Fernando WJN. Solid heterogeneous catalysts for transesterification of triglycerides with methanol: A review. Applied Catalysis A: General. 2009;**363**:1

[107] Gerpen JV. Biodiesel processing and production. Fuel Processing Technology. 2005;**86**:1097

Chapter 5

Nanocatalyst Mediated Biodiesel Production from Waste Lipid as Feedstock: A Review

R. Dayana, P. Bharathi and G.M. Shanthini

Abstract

Petroleum-based fuels are widely utilized and pose a threat to the environment, necessitating an urge to bring up an equally effective substitute. Globally, research is focused on biofuel production from various sources which is renewable, highly affordable, and has lesser carbon emission. Biomass is used as raw material to produce biodiesel to achieve clean, green, and renewable fuel. Edible and nonedible raw materials are utilized for the production of biodiesel. Biodiesel from lipid sources produced through the transesterification process serves as an effective alternative for the production of renewable fuel with reduced carbon emissions and greenhouse gases. The cost of biodiesel is dependent on raw materials and catalysts. The acidic and basic homogeneous catalysis reaction has a corrosive effect during synthesis and poses a risk in scalability. The heterogeneous reaction is costlier and has poor performance in the transesterification of lipids. Raw material contributes to 70–80% of the overall production cost. Municipal sewage sludge (MSS) is rich in lipid content and serves as promising raw material for biodiesel production. Nanocatalyst has superior activity in producing pure products with fewer side reactions. This paper reviews the lipid extraction techniques and biodiesel production from MSS using various nanocatalysts.

Keywords: biodiesel, municipal sewage sludge, nanocatalyst, transesterification, lipid extraction

1. Introduction

Petroleum-based fuels contribute a significant impact on the business economy in developing countries toward various applications such that the transport of goods from industries and agricultural products, in operating diesel tractors and pump sets in the agricultural lands. Economic growth is associated with the rate of transportation. The energy demand is always incorporated with the industrialized world and domestic sector. Due to high energy demand, the fossil fuel necessity also increases which leads to a large amount of pollution, hence it is necessary to develop renewable energy sources. Therefore, it is a stipulated time to focus on alternative sources. Mainly focus on alternative fuels should have technically economically feasible, smaller environmental impact, and be readily available [1]. The MSS is one of the high lipid sources, it is obtained mostly from the domestic sector which includes long-chain fatty acids, grease, and fats, and an interesting factor is the microorganisms contain phospholipids present in their cell membranes, and during cell lysis, some of the byproducts and their metabolites acted as a major lipid source in MSS. This could be substantiated by the research work that has indicated that the lipid content in the sewage could be a conceivable feedstock for biodiesel production [2].

Biodiesel could be one of the solutions to overcome the issues of petroleum-based fuels, as, it is a fresh and renewed form of energy and it could own the worldwide markets due to its non-toxicity, biodegradability, environmentally beneficial, and similar ignition characteristics to fossil fuels [3]. Lower emissions of carbon monoxide, sulfur, and other hydrocarbons make biodiesel a carbon-neutral fuel [4]. It is a sustainable energy source used to reduce global warming [5]. Studies and recent research works revealed that using biodiesel can minimize CO_2 release to 78% which helps to prevent nature from pollution [6]. With the help of non-edible feedstocks, India produces 10–250,000 tons of biodiesel per year. According to the survey of 2014, the overall consumption of biodiesel is 1000 barrels/day. The usage of biodiesel economically supports the growth of India's early $1.47 billion foreign currency. Rapid growth in Industrial sectors and transport sectors are causes of polluting environment. Renewable energy sources are the best way to overcome this environmental crisis. Biodiesel production is one of them which supports technically and economically to the society.

2. Why biodiesel?

India's growing economy, multiplying population, and rapid urbanization result in increasing energy demand. Accordance to the report from International Energy Agency (IEA), the energy requirement is more due to the high depletion of energy in the past few decades. The IEA report says that a 6 million barrels a day increase in oil consumption results in major energy demand in India. The rise in energy demand is mainly due to the increase in individual vehicle ownership. The planning commission's 2002 statement is that "The primary commercial energy demand has grown at the rate of 6% between the years 1981-2001". On the other hand, India is facing a coal shortage of 23.96 million tons in the past few decades due to the depletion of energy sources and an increase in energy demand. For a decade of years, the demand for natural gas is increasing at the rate of about 6.5%. Renewable energy sources offer an alternate supply of energy to face the energy crisis. India is the highest potential for renewable energy sources. The significant renewable energy sources are wind energy, small hydrothermal energy, solar energy, and biomass. Though India produces more renewable forms of energy, the energy dependence is expected to increase further by 8% to achieve the gross domestic product growth rate of the tenth five-year plan. To solve this issue Government of India has agreed on a high priority for the energy sector and renewable energy resources. Tamil Nadu is a pioneer state in promoting the production of renewable energy resources. At the Indian level, Tamil Nadu contributes 27% (29989.21 MW) of total renewable energy via renewable energy sources. There is a steady increase in non-conventional energy sources in the past 10 years. The main contributors of non-conventional energy sources are wind energy (61%), biomass (35%), and other renewable forms of energy contributing to the remaining energy sources. Tamil Nadu is facing a major power crisis as Tangedco (The state power utility) is running short of coal. Tangedco imported 1.4 million tons of coal by the year 2018. Due to climatic changes and cyclonic storms in the Arabian Sea, there is a drop in wind energy generation on the south side of Tamil Nadu. Therefore, there is a high demand for energy supply in the state as coal and non-conventional energy sources are kept on depleted.

3. Municipal sewage sludge (MSS) as a promised raw material

Sewage is untreated municipal waste generated by domestic, industrial, and commercial sources. The level of wastewater production is increased rapidly due to the population, economic development, urbanization, and improving living conditions. Central Pollution Control Board (CPCB) surveys depict that "There are a total of 269 sewage treatment plants in India of which 213 are in proper working condition. In urban centers of the nation, the total sewage generated per day is about 38 billion liters a day but the total sewage treatment plant capacity is around 12 billion liters a day". The Ministry of New and Renewable Energy (MNRE) suggests that there is a probability to produce nearly 226 MW of energy from sewage sludge of treated and untreated sewage sludge sources. Sewage sludge is the waste generated in huge amounts in the waste treatment processes after primary and secondary treatment processes [7]. Also, the sludge is rich in lipid content as the surface of the sludge is proficient enough to adsorb different forms of lipids [8].

4. Treatment methodologies of MSS

In the sewage treatment plant, there are mainly 4 steps involved in the treatment of sewage sludge (**Figure 1**). The main 4 steps are

a. Pretreatment and grit removal

b. Primary treatment

c. Secondary treatment

d. Tertiary treatment

In the pretreatment process, easily removable materials like waste material, liters from trees, and large-sized materials will be removed from the raw sewage. The influent is passed into the bar screen where larger objects like plastic packets, liters, and cansticks present in the sludge were removed. Grit is composed of sand, gravel, heavy materials, organic material, fine solid particles, etc. The grit removal is done for the treatment of equipment with closely machined metal surfaces such as commuters, fine screens, centrifuges, heat exchangers, and high-pressure pumps. A grit removal system is present in sewage treatment plants for reasons which include; reducing the necessity for frequent cleaning of digester due to the accumulation of grit and lessening the settling of grit in the treatment tanks and passage pipes. The removal of grit also protects the digesters and clarifiers from wear and abrasion. The sewage flows through large tanks called "pre-settling basins" in primary sedimentation. This leads to a rise in the level of grease and oil on the surface of these tanks due to the continuous agitation of the sludge. The primary settling tanks have a scrapper and agitation tank and they continuously direct the collected sludge toward the sludge treatment facilities. After the primary clarification, the primary sludge is collected. The primary sludge is the combined form of free-floating grease and settled solid matter [9]. The secondary treatment involves the degradation of the biological content which could be human/food waste, and the removal of the organic content and the suspended solids which passes the primary treatment. In the secondary treatment, the settled

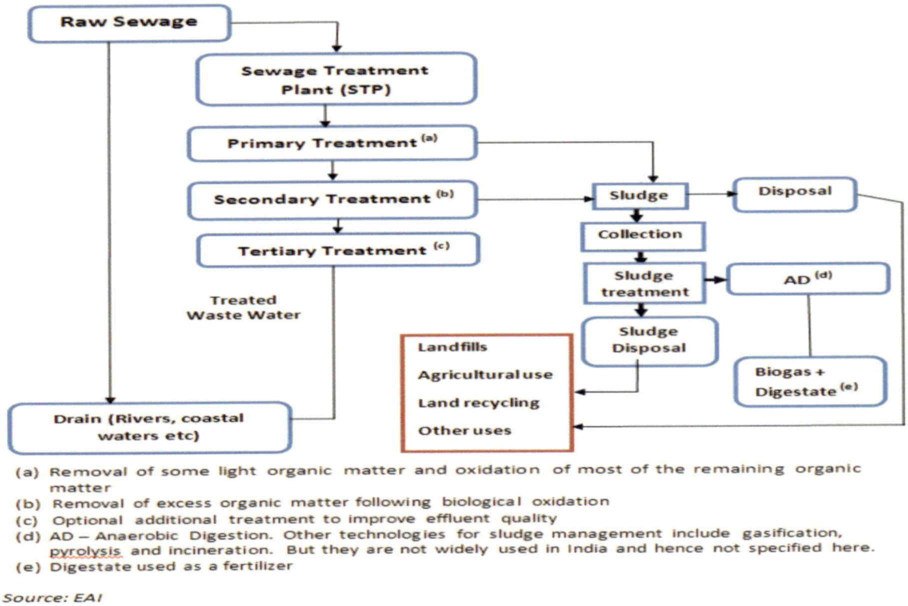

Figure 1.
Overall view of sewage treatment plant.

sewage liquor is treated with an aerobic biological process. The secondary treatment methods use filtration and aerobic treatments to separate and break down the content received in the secondary treatment unit [10]. The secondary sludge primarily encompasses microbial biomass and settled solids produced during the aerobic biological treatment of primary treated wastewater. Thus, the secondary sludge comprises the lipids from the lysed cell with lesser free fatty acids compared to the primary sludge [11]. The final stage of treatment is tertiary treatment. It is also known as effluent polishing. It is done to improve the quality of effluent when discharged into environments such as the sea, rivers, and lakes. The eradication of chemical pollutants obtained from pharmaceutical industries, particles present in household chemicals, chemical effluents for small-scale industry, and agricultural pesticides is quite difficult in the conventional sludge treatment procedures hence these waste/effluents might pollute the water bodies in the disposal areas. In spite of the lesser concentration of the disposal, these pollutants are sufficient enough to be toxic to aquatic life. Also, pharmaceutical disposals that could induce genotoxicity and microbial resistance are considered to be toxicologically relevant pharmaceutical pollutants. The reduction of the fourth stage of treatment is being followed in many countries recently. Additional to this odor control, biological nutrient removal, phosphorous removal, nitrogen removal, and disinfectants were also added before disposal.

5. Production methodology of biodiesel from MSS

The primary sludge flocs and secondary sludge has utilized as efficient raw materials for biodiesel production via the transesterification process. Initially, the lipids are extracted using various solvents. The stagewise extraction process was used to separation of lipid content from MSS. The efficiency of lipid extraction depends on the types of solvents or mixed solvents and the number of stages. Bharathi and Pennarasi [12] reported that using various solvents and their extraction efficiency of lipids from MSS (**Table 1**).

Lipids extraction is followed by the transesterification process for the production of biodiesel It is a chemically Fatty Acid Methyl Ester (FAME), formed by the transesterification reaction (**Figure 2**). Transesterification is a process that involves the swapping of the R (alkyl) group of triglyceride's esters with the R′ (alkyl) group of the alcohol compound [1].

The transesterification reaction may be a catalytic or non-catalytic process. Among these, the noncatalytic reactions are very slow and result in lesser yield compared to catalytic reactions. Hence, the transesterification reaction is manifested with the help of the catalysts and is a high-yielding process. Briefly, the complete transesterification process includes:

i. conversion of triglycerides to diglycerides,

ii. conversion of diglycerides to mono-glycerides and ultimately,

iii. conversion of mono-glycerides to glycerol along with the formation of biodiesel

6. Prominent status of nanotechnology

Nanotechnology is a wide field in which nano-sized materials are used for various applications. Nano-sized materials are in the size of 10^{-9} nm. Such particles are said to be nanoparticles. The nanoscale ranges from 1 to 100 nm (**Figure 3**). The advantage of nanoparticles is the reactive area gets increased as the size reduces from bulk size. This has been used in the past few decades in the form of silver and gold glittering paints in the church, palaces, etc. Such old technology has been tuned even better to increase the application in all the emerging fields. There are various things in nature in nano size, apart from that; mankind played a major role in creating new nanomaterials for their relevant applications.

There are a lot of industries and researchers concentrated on nanotechnology and its applications (**Figure 4**). They explore the technology in various fields like improvement in energy efficiency, telecommunication fields, cosmetics, textile industry, foods, medicines, etc. The efficiency of fuel production is considerably improved using nanotechnology. One of the fuel production methods is from natural materials and low-grade

Solvents	Lipids extracted from primary sludge (g)		Total (g)	Lipids extracted from secondary sludge (g)		Total (g)
	Stage1	Stage2		Stage1	Stage2	
Chloroform-Methanol	2.75	0.50	3.25	1.90	0.55	2.45
N- Hexane	1.83	0.62	2.45	1.65	0.45	2.00
Diethyl ether	1.72	0.60	2.32	1.70	0.45	2.15
Ethanol	1.10	0.40	1.50	1.26	0.50	1.76

Table 1.
Quantity of lipids extracted from MSS [12].

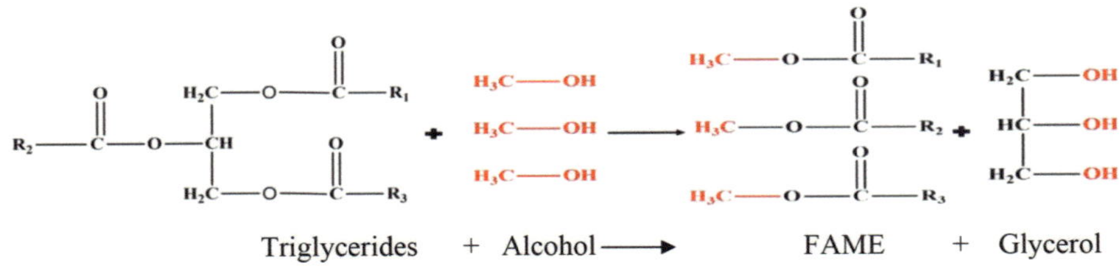

Figure 2.
General equation of transesterification of triglycerides [12].

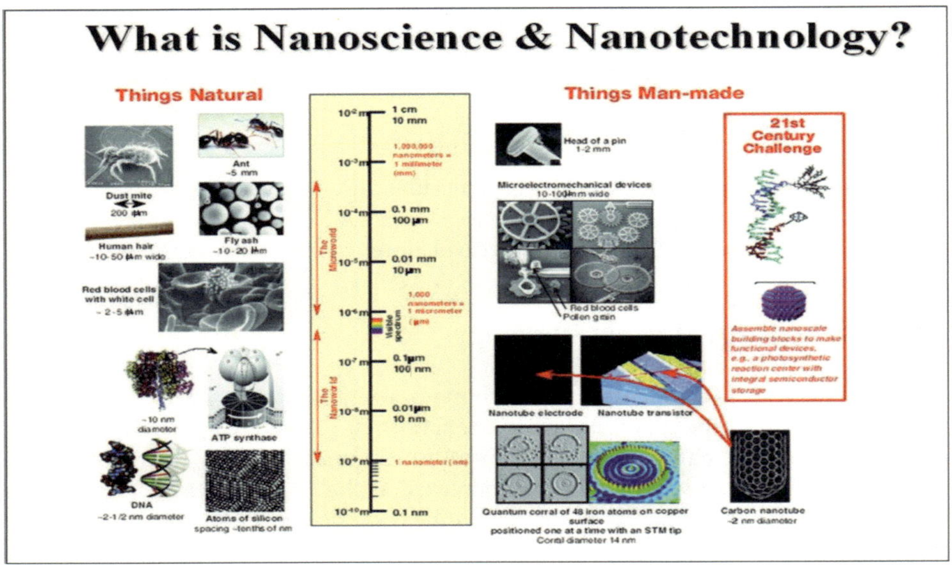

Figure 3.
Natural and artificially produced nano-sized things (https://philebersole.files.wordpress.com/2012/04/nanotechnology.jpg).

crude oil by improving catalysis. Through nanotechnology, the fuel efficiency in vehicles and power plants is improved through friction, which increases efficiency and reduces combustion. Nanotechnology plays a vital role in telecommunication transceivers. The nanoscale-based transceivers reduce the system complexity and improve the Quality of Service (QoS). Faster and more power-efficient electronic devices are constructed using nanoscale transistors and diodes. Nowadays, nanoscale type memory devices are used to equip supercomputers eg. Magnetic Random Access Memory (MRAM), which enabled tunnel junctions of magnetic nanoparticles, which quickly and effectively saves the data during a system shutdown or crash. Such nano-based chronic devices are used in aircraft communication as well as the resumption of play and used for the collection of accident data vehicles. Advancements in various cosmetic products like lotions, dermatological creams, shampoos, sunscreen lotions, and specialized makeup products are prepared by superimposed nanomaterials. In the cosmetic field, nanomaterials provide greater clarity and good coverage in cleansing, absorption, customization, antioxidant, and anti-microbial. Nanotechnology not only plays the role in the cosmetic industry but also in the food industry. Nano-engineered nanocomposite materials support food packaging in the food industry. Food packaging reduces the escape of carbon dioxide from the

Figure 4.
Applications of nanotechnology (https://www.researchgate.net/profile/Karolina_Niska/publication/317719241/figure/fig 2/AS:555489508900864@1509450427466/Applications-of-nanotechnology.png).

soda and reduces the amount of oxygen, humidity, or the growth of bacteria to keep fresh and safest food for a longer period. Nanosensors embedded in plastic containers warn of spoiled food and are currently being developed nanosensors for the detection of pesticides and other contaminants in food before packaging and distribution. The use of nanotechnology in the medical field has vast appliances. It is used to stimulate the neurological systems and their growth. Nano-sensors are used to predict the damage to spinal and brain tissues. One of the methods supports structured nano gel filling the space between existing cells and encouraging new cells' growth. Quantum dots are semiconductor nanocrystals that can improve biological imaging for medical diagnosis. Researchers are developing customized nanoparticles that can deliver drugs directly to diseased cells in the body. When it is perfected, this method greatly reduces the damage during the treatment (eg: chemotherapy) done to the patient's healthy cells. Marvelous support of nanotechnology to the environment is to meet the need for clean drinking water which is affordable through low-cost rapid detection of impurities, purification, and water treatment systems. For example, researchers have discovered unexpected magnetic interactions between the very small spots that can help in the removal of arsenic or carbon tetrachloride from water and are currently being developed filters that can remove virus cells from nanostructured water and they were investigating how the ionization electrode using nano-sized fibers reduce the energy requirements and costs of removing salts from water. It is also used to degrade solid waste with less time when compared to the naturally occurring degrading process. It is also helpful in the process of conversion of waste sources to renewable energy available for use. Manufacturing the fabric with nanoparticles allows the introduction of upgraded fabric properties without a substantial increase in weight, thickness, or stiffness, unlike the other modification techniques. The photocatalytic activity of the nano TiO_2 in the fabric treated with Nano TiO_2 could impart an anti-bacterial effect and might reduce the staining. Nanoparticles are primarily used in the catalysis of chemical reactions. Hence, the usage of nanoparticles possibly reduces the need for additional catalysts that are necessary to achieve the desired results, with lesser capital and pollutants. The main applications involved in the field of oil refining, catalytic converters in cars, and catalytic reactions in renewable bioenergy production. The functional nanomaterials with antimicrobial properties can be used to build high-performance systems on a small-scale or point of use to increase the robustness of the network water supply and water networks that are not connected to the center and the emergency response network after catastrophic events. Nanotechnology can also be applied in the detection of even the minute quantity of gas and vapors with the highest sensitivity. Nanomaterials, such as quantum dots, carbon nanostructures, metal-based nanomaterials, and metal oxide semiconducting nanomaterials can be used as the gas sensing elements in the nanosensors. The nanoscale dimension of these nanomaterials helps in sensing the lesser level of the gas molecules that are adequate enough to tune the electrical properties of the detecting element, allowing the detection of the tiny concentration of gas vapors when nanoparticles are used. Nanotechnology can enhance the catalyst performance in transforming the vapor released from automobiles or industrial plants into harmless gasses as nanoparticles have a larger surface area which improves the interaction with the reacting chemicals compared to traditional catalysts. In addition, the larger surface area allows more chemicals to interact with the catalyst, making them ideal for sensing with higher sensitivity.

7. Role of nanocatalysts in various fields

The nanocatalysts have their own property which was usually targeted for specific applications. The four major properties are

- Magnetic property
- Optical property
- Electronic property
- Catalytic property

Nanocatalysts have placed their impact on almost all fields due to their multiple properties and uses toward the problem, thereby creating the best solution for the problems. Nanoscience contributes to the reduction in the size of devices and technologies with upgraded properties. The key applications of nanotechnology include, water purification, energy storage devices, biodiesel production, medical applications, dye reduction, fuel cell, carbon nanotubes, etc. [13].

Nanocatalysts are widely used in the following fields for a huge number of applications as follows:

- Chemical sensors
- Better air quality
- Electronics
- Sporting goods
- Batteries
- Fuel cells

- Solar cells
- Catalysis
- Medicine
- Textile industry
- Environment
- Water treatment
- Cosmetics
- Fuels
- Energy
- Food
- Space

Nanocatalyst is used against environmental pollution like wastewater treatment, soil remediation, waste degradation, chemical pesticide degradation etc. Nanoparticles are used as a catalyst in biodiesel production and plays the main role in photodegradation of methylene blue. Nanoparticles have applications in soil-plant systems in many ways like the delivery of pesticides and biopesticides, pesticide degradation, nanosensors for plant-pathogen detection, and also in fertilizer-controlled delivery [14]. Nanoparticles are employed in the delivery of genetic materials, in plant protection and nutrition like soil remediation, slow release of fertilizers, pesticide degradation, and in seed treatments [15]. The main important application of nanoparticles in agriculture was nanofertilizers. Nanofertilizers improve crop growth, yield, quality, and reduce fertilizer usage and cost for cultivation. It increases plant growth mainly by increasing the photosynthesis rate and also prevents plants from abiotic and biotic stress [16]. Applications of nanotechnology in precision agriculture are nano biosensors, early detection of viral disease in plants, nanoparticles serving as micronutrients for plants, nano herbicides, nano fungicides, and in insect pest management [17].

8. Nanocatalyzed transesterification process

Nanotechnology has an increasing impact in the fields of biotechnology, pharmaceutical technology, and pure technological applications. Nanotechnology places its main role in the conversion of biomass to bio-energy in various renewable resources. Implementing nanotechnology in the bioenergy production process gives a high impact on bio-energy research. Four categories of nanoscale area units are investigated as useful materials for water purification which are metal-containing nanoparticles, carbon nanostructures, zeolites, and dendrimers [18]. Nanotechnology reveals good results than the other techniques used in water treatment as they exhibit higher interaction with high surface area (surface/volume ratio) [19]. The present investigation of the potential applications of nanotechnology in water and wastewater treatments are adsorption, membrane processes, photocatalysis, antimicrobial efficiency, sensing, and monitoring [20]. Copper-doped zinc oxide nanocomposite (CZO) comes under heterogeneous catalyst as it contains both copper and zinc oxide nanoparticles in it. It shows a positive sign in the field of emerging catalysts as they are non-corrosive and can be easily separated from the product mixture. Doping is significantly used in enhancing the optical and electrical properties of semiconductors. Doping with transition metals leads to improving the interesting properties of zinc oxide [1]. CZO nanocomposite comes under heterogeneous nanoparticles, it can also be recycled and reused. CZO nanocomposite is a bifunctional heterogeneous catalyst. They gain attention because they carry out transesterification and esterification processes for both acid and base-containing reactions. The use of nanoparticles/nanocomposites as catalysts provides higher catalytic activity and selectivity due to its nano-dimension and morphological structure. Thus, the CZO nanocomposite can be used as a catalyst for the transesterification of lipids obtained from municipal primary sewage sludge. It increases the yield of biodiesel and can be recovered, recycled, and reused [21]. After the transesterification process, the mixture was separated and the byproducts are separated. The biodiesel had to be subjected to washing subsequent to the transesterification process to eliminate excess catalysts, methanol, glycerol, and soap. After washing, the biodiesel sample must be heated to evaporate solvents and water present in the mixture [22]. Gas chromatography-mass spectrometry (GC-MS combines the features of gas chromatography and mass spectroscopy to identify the different substances present in the test sample. Applications of GC-MS include drug discovery, fire investigation, environmental check, explosives examination, and documentation of unknown samples. In addition, it can help in the identification of the trace elements in the samples which were earlier assumed to have disintegrated beyond identification. GC-MS is broadly used in the analysis of the compounds such as esters, fatty acids, alcohols, aldehydes, terpenes, etc. The below process flow chart clearly shows the production of biodiesel from MSS via the nano-catalyst transesterification process (**Figure 5**).

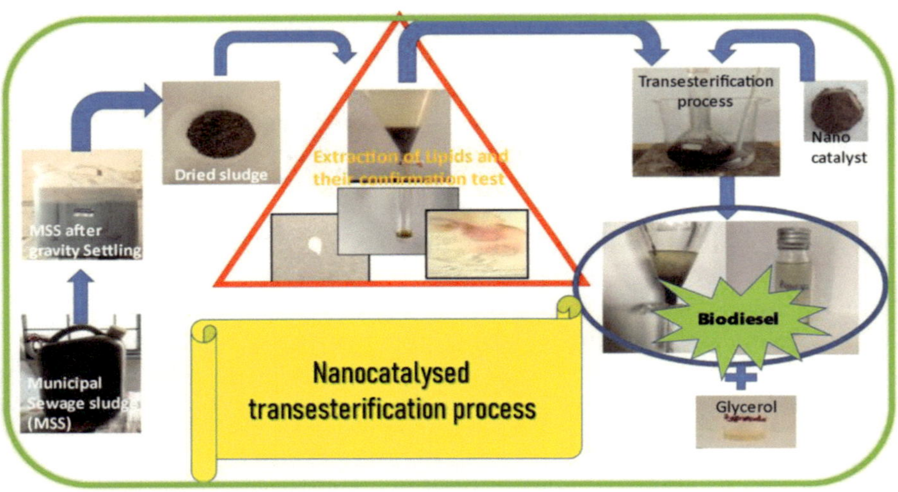

Figure 5.
Process flow chart of biodiesel production from MSS.

9. Quality standards of biodiesel

Quality is meant for long time goals, exactly successful use, without any technical problems of biofuel. It can be depending on many factors: mainly, the quality of the raw material chosen, lipid content composition. The physicochemical properties of biodiesel were the most significant parameters to speak about the quality of biodiesel. Some of the important physical and chemical properties are listed in **Table 2** with their ASTM standards (ASTM D6751-06).

These quality standards of biodiesel on the market are affected by various factors, which vary from place to place, and country to country. Mainly, it compares the characteristics of the existing diesel fuel standards, types of diesel engines commonly used in practice region and their emissions, and climatic properties around the regions. Therefore, the quality standards of biodiesel are varied on country/region. Not surprising that there are some considerable differences between the regional standards. **Table 3** shows a worldwide important biodiesel quality standard.

Notable properties of biodiesel

i. Cetane number (CN)

It is used to measure the combustion properties of diesel fuel. It is a measurement of the quality of fuel. Modern highway diesel engines require a CN ranging from 45 to

Properties	SI Unit	ASTM limits	ASTM methods
Physical			
Flash Point	°C	7130 min	D93
Density	kg/m^3	N/A	D4052
Kinematic viscosity	mm^2/s	1.9–6.0	D445
Water & sediment	Vol%	0.050 MAX	D2709
Cloud point	°C	N/A	D2500
Pour point	°C	N/A	D975
Distillation temperature	°C	360	D1160
Chemical			
Acid value	Mg KOH/g	0.8	0.664
Cetane No.	—	47 min	D613
Free glycerin	Mass%	0.020	D6584
Sulfated ash	Mass%	0.240	D6584
Carbon residue	Mass%	0.05	D4530
Sulfur	Mass%	0.05	D5453
Copper strip corrosion	—	3 Nos. max.	D130
Phosphorus	Mass%	0.001	D4951
Oxidation stability	Hours	3	EN15751
Methanol content	Mass%	0.2	EN14110

Min – minimum, Max – maximum, N/A – Not available.

Table 2.
Physical and chemical properties of biodiesel compared with ASTM standards (ASTM D6751-06).

Country/Area	Specifications	Title
EU	EN 14213	Heating fuels - Fatty acid methyl esters (FAME) - Requirements and test methods
EU	EN 14214	Automotive fuels - Fatty acid methyl esters (FAME) for diesel engines - Requirements and test methods
The U.S.	ASTM D 6751 -11a	Standard Specification for Biodiesel Fuel Blend Stock (B100) for Middle Distillate Fuels
Australia	—	Fuel Standard (Biodiesel) Determination 2003
India	IS 15607	Bio-diesel (B 100) blend stock for diesel fuel - Specification
Japan	JASO M360	Automotive fuel - Fatty acid methyl ester (FAME) as a blend stock
South Africa	SANS 1935	Automotive biodiesel fuel

Table 3.
Biodiesel standards [23].

55. The higher CN, the more fuel burns better within the engine. And also, it is associated with ignition delay time, engine knock, carbon monoxide emissions, combustion efficiency, smoothness, and hydrocarbon and nitrous oxide emissions [24, 25]. The length of the hydrocarbon chain increases with increasing CN, which leads to a decrease in the saturation state in the carbon chain of the Fames. The estimation of the CN formula was given below for individual fatty acids. These models were developed using a combination of [26, 27].

$$CN = 1.068\ \Sigma\ (CN_i m_i) - 6.747$$

Where CNi – Cetane number, m_i – The mass percentage of FAME

ii. Viscosity

It is the predominant parameter with an impact on the fuel quality, since it plays a vital role in the fuel atomization process. High viscosity can lead to the coking process, and affect the combustion and emission rates from the engine. Generally, viscosity decreases as temperature decreases which leads to fuel components being saturated and causing precipitate problems. This may arise from clogging the fuel lines, pumps, and filters [25, 26]. As per the EN14214 guidelines, the specification for viscosity is 3.5–5 mm^2s^{-1} for 40°C. The degree of saturation of compounds depends on the viscosity of the fuel [28].

Allen et al. [29] proposed a mixture model for the estimation of viscosity,

$$\ln \mu_m = \sum_{i=1}^{n} y_i \ln \mu_i.$$

where μ_m is the viscosity of the mixture and Yi and μ_i mole fraction and viscosity of individual component I respectively.

iii. Oxidation properties

It is a difficult parameter to understand properly not unlike other properties such as cetane number since it is dependent on the bulk composition of the fuel. The double bonds present in the Fames are prone to autoxidation, which leads to deterioration problems in the fuel system [28]. It is used to measure the fuel resists oxidative degradation property which determines the fatty acids occurrence with double bonds in feedstocks. A non-linear relationship between oxidation rate increases with an increasing double bond in a Fame. Minor components provide disproportionate effects on oxidative stability inhibiting oxidation. Therefore, to change the oxidative stability, any of the other properties to be improved by additives.

iv. Density

It is a significant parameter, with an impact on fuel quality. The density of biodiesel usually varies between 0.86 and 0.90 g/cm^3 [30]. The non-edible biodiesel densities are also in the range of 800–965 kg/cm^3 [31]. Generally, molecular weight is one of the important factors that contribute to fuel density. Densities are calculated with respect to various temperatures, instead of the standard value at 15°C, available at EN 14214. The ASTM standard procedures are also available at 15°C, required by D1298 [31]. The liquid densities are calculated by following the formula using the relationship between the molecular weight and molar volume.

$$\rho = \Sigma\, x_i\, MW_i/\Sigma x_i\, V_i.$$

where x_i is the molar fraction of the component and Vi is the molar volume of a liquid [32].

v. Flashpoint

It is the lesser temperature at which the liquid gives off its vapor, the lower the flash point the easier to ignite the material. Mostly the flash point of the combustible liquid is 104°C. It is used to analyze the hazards of fuels, which are less than 37.8°C are called flammable, whereas above that temperature are called combustible. Each biodiesel has its own flash point. Moreover, the flash point is affected by many factors including the number of double bonds, the number of carbon atoms, the chemical composition of the biodiesel, pressure, oxidant, and apparatus sheltering [33]. Different test methods are available to determine the flash point of biodiesel which includes the EN test method and the ASTM test method. In addition to the observation that both boiling point and flashpoint decrease with decreasing pressure, the literature, and the ATSM standard D6751 recommend a flash point of biodiesel is 130°C except for non-edible biodiesel.

vi. Cloud point and pour point

It is the minimum temperature at which the first nuclei crystal formation takes place. The pour point is the minimum temperature below which a liquid loses its flow characteristics. Comparatively, the cloud point is of a high value with high temperature and a low value with low temperature. Generally, biodiesel has higher cloud and pour points than conventional fuel [34]. Specifically, the standard ASTM D2500 and D97 test methods are available to measure the cloud and pour point. Generally, most biodiesel properties are matched with diesel fuel except cloud and pour point. These are low-temperature properties. Much literature indicated that the cloud point of pure biodiesel is around 13°C, and for pure biodiesel is 0°C. Cloud points may be achieved by adding liquid additives. It has a very low solidifying temperature and is highly soluble in biodiesel.

10. Economic impact of biodiesel

Energy consumption is of great concern globally due to its availability. Also, global energy consumption is increasing day by day. To IEO 2016 projection, over the years 2010 to 2040, the total energy consumption is expected to increase by 50% compared to previous decades. The major part of the energy is being utilized by the developed countries due to their richness in economy and population. Apart from energy sources like coal, natural gas, and crude oil; petroleum-based fuels are in great demand. These fossil fuels are scarce and impose a negative impact such as the production of greenhouse gases and causes air pollution. Also, the excessive utilization of these fossil fuels is heading toward the global depletion of natural resources, demanding an equal and effective alternative, giving birth to renewable energy resources such as biodiesel. Biodiesel is generally produced by trans-esterifying fat-based feedstocks like waste cooking oil, oil-based industrial effluents, vegetable oil, and animal fat, waste sludge. The cost of the final product (biodiesel) will vary according to the kind of feedstock. Shifting fossil fuel to biodiesel is not easier due to less yield and high production cost. Researchers continued exploring the various options of biodiesel synthesis to reduce production costs, as it poses a major challenge in the extensive utilization of biodiesel.

Considering the traditional biodiesel etproduction technique, the most important costs in the production of biodiesel are capital investment costs and operating costs. The costs covered under capital investment are equipment and plant establishment costs. The plant establishment includes the cost of installation of equipment, instrumentation, pipelines, electrical lines, and a few other auxiliary developments. The operation costs include feedstock, catalyst, utility, labor charge, maintenance, and repair of equipment [35, 36]. Among these expenses, the feedstock significantly impacts biodiesel's economy, imparting more than 75% of the total production cost [35]. To reduce the feedstock cost, the poor quality of feedstock or reuse of feedstock is much preferred. These will influence the excessive release of free fatty acids (FFA) and increase the water content. The presence of excess FFA and water content in turn reduces the yield and quality of biodiesel [37]. To improve the yield and quality, additional processing steps should be carried out which in turn increases the production cost. A better and good quality of feedstock without compromising the food demand and environmental impact and which could reduce the final cost is much preferred and being in trials by many researchers and scientists. The choice of feedstock varies country-wise by their utilization of crops. Argentina prefers soybean as a feedstock owing to its less cost. While China does not choose soybean as a biodiesel feedstock as soybean oil is a staple demand in Chinese foods [38]. Jatropha and castor oil-based feedstocks are favorites in India [39]. Waste cooking oil and animal fats are ideal in countries like Japan, Canada, and Australia [40].

Waste cooking oil is one of the better preferences of feedstock in producing biodiesel. By utilizing waste cooking oil, roughly 1.5 ratios of energy output to input, can be obtained and approximately two times income can be obtained compared to the total expenses [41]. A study on biodiesel production from *Calophyllum inophyllum* oil gave an idea of the economic performance of heterogeneous catalyst-based production of biodiesel. The influence of the cost of feedstock on the net present value and payback time was investigated in detail. The cost of the feedstock is in the range of 0.2–0.5 $/kg. In the case of a lesser feedstock price of 0.2 $/kg, the net present value is about 31 million dollars, with a payback time of 0.41 years. While the feedstock purchase cost of 0.5 $/kg and above poses a negative impact on net present value with the payback period being the time past the project completion time. Apart from lesser feedstock cost, the recovery and utilization of methanol and heterogeneous nanocatalyst could greatly reduce the cost of biodiesel [42]. The non-edible source of feedstock like microalgae is used to reduce the exploitation of food crops and oils in the production of biofuels. Microalgae as a feedstock are advantageous concerning their higher production rate, lesser environmental impact, and lesser requirement of land for setting up a biodiesel plant, which could in turn reduce the cost of biodiesel. In recent years, microalgae have been one of the most promising feedstock choices to produce biofuel which could replace approximately 50% of the fossil fuel demand as the mass cultivation of microalgae and its availability could greatly bring down the production price of biodiesel [43].

In addition, the cost of biodiesel varies greatly, in accordance with the type of catalyst used, which can be either homogeneous or heterogeneous. A heterogeneous catalyst has the limitations like deprived interaction as there might be restricted diffusion of the catalyst when it exists in different phases of reagents. In the case of a homogeneous catalyst, the separation and reuse of the catalyst are quite tedious. The behavior of nanocatalysts lies in between homogeneous and heterogeneous catalysts, with notable advantages like good selectivity, sensitivity, separation, and reuse of catalyst [44, 45]. The careful selection of catalyst reduces the number of processing steps and favors the reusability of the catalyst, which in turn supports minimizing the total production cost of biodiesel [46]. The magnetic nanomaterial-based catalyst is proven to be very effective in biodiesel production compared to the conventional catalyst, owing to its availability, size, surface area, high surface-to-volume ratio, stability to different reaction conditions, resistance to saponification, reusability, and wide synthesis options [46]. Also, the higher surface-to-volume ratio of the nanocatalyst favors simultaneous reactions to take place at the same time which in turn accelerates the process [47]. The acceleration of the reaction process contributes to minimizing the production cost. In addition, magnetically separable nanomaterials can easily eliminate the separation protocol for the recovery of catalysts [23, 48]. Bharathi and Pennarasi [12] estimated the cost of sludge biodiesel (**Table 4**). The overall production cost is 3.11 USD per gallon, but they did not include the byproduct (glycerol) sale cost. Surely it will reduce the overall biodiesel production cost as well.

Particulars	Cost per gallon (US $)
Centrifuge O&M	0.43
Drying O&M	1.29
Extraction O&M	0.34
Biodiesel processing O&M	0.60
Labor	0.10
Insurance	0.03
Tax	0.02
Depreciation	0.12
Capital P&I service	0.18
Total cost	3.11

Table 4.
Cost estimation for sludge biodiesel [39].

11. Summary

Biodiesel has several benefits over diesel fuel. The production cost only is a major problem compared to petro-diesel. And also, vegetable oil and animal fats are the raw materials for biodiesel production causing the cost of biodiesel to increase. MSS is a waste and lipid source readily available for the extraction of lipids for biodiesel production. MSS treatment for biodiesel production may be used in a waste treatment facility with limited expenditure on raw materials. Also, municipal waste might not contribute to the landfill issue and renewable sewage waste can be considered for biodiesel production instead of depending on edible sources of raw material. But there are some disputes with MSS. The pre-treatment process includes collection, dewatering, drying, etc. The lipid extraction steps also need more amount of organic solvents which will increase the production cost. However, the amount of lipids depends on the type of sludge. Sludge-to-solvent ratio, extraction time, and solvent recovery are the factors that affect the efficiency of lipid extraction and cost. Optimization is only one solution to reduce the cost concerning the above factors. The role of catalyst for biodiesel production is also the most important. Acid and alkali transesterifications are high-cost and slow processes. Comparatively nanocatalyst transesterification is less costly and fast. Factors such as the choice of nanocatalyst based on its properties and type of feedstock will be highly favorable in improving the transesterification reaction. Based on the demand of the biodiesel market, the designing of these factors can be modified and improved. The nanocatalysts which are being developed in the recent past are highly preferred for the heterogeneous catalysis reaction as they are highly suitable for improving the efficiency of the transesterification process. Nanocatalysts can play a major role in improving the yield of biodiesel as they possess good surface area which could favor the catalytic efficiency and trigger the overall catalytic reaction. In spite of all these advantages, extensive research work is necessary to get optimization for biodiesel production from MSS and to study the toxicity levels of nanocatalysts before utilizing them in biofuel production and making a good profit from any source of raw material. Nanocatalysts could serve as efficient catalysts as they are recyclable, have less production cost, and possess a good life which could greatly influence the overall production cost of biodiesel. It is observed that the primary energy demand is set to double by 2040, therefore the use of alternative fuels such as biodiesel is bound to grow as well. In the meantime, it is observed that the percentage of MSS also increased due to the high population. This review insists on the methodology available for the production of biodiesel from MSS. It strongly helps future research thereby there is one forward step to save the environment and increase the global economy values.

Author details

R. Dayana[1], P. Bharathi[2*] and G.M. Shanthini[3]

1 SRM Institute of Science and Technology, Chennai, India

2 Rajalakshmi Engineering College, Chennai, India

3 Bharath Institute of Higher Education and Research, Chennai, India

*Address all correspondence to: bharathipurush@gmail.com

References

[1] Meher LC, Vidhya SD, Naik SN. Technical aspects of biodiesel production by transesterification. Renewable and Sustainable Energy Reviews. 2006;**3**:1-21

[2] Girisha ST, Ravikumar K, Mrunalini GV. Lipid extraction for biodiesel production from municipal sewage water sludge. European Journal of Experimental Biology. 2014;**4**(1):242-249

[3] Arabani A. Non-edible vegetable oils: A critical evaluation of oil extraction, fatty acid compositions, biodiesel production, characteristics, engine performance, and emissions production. Renewable and Sustainable Energy Reviews. 2013;**18**:211-245

[4] Miller P, Kumar A. Development of emission parameters and net energy ratio for renewable diesel from canola and Camelina. Energy. 2013;**58**:426-437

[5] Dias APS, Bernardo J, Felizardo P, Correia MJN. Biodiesel production by soybean oil methanolysisover SrO/MgO catalysts: The relevance of the catalyst granulometry. Fuel Processing Technology. 2012;**102**:146-155

[6] Tabatabaie SMH, Tahami H, Murthy GS. A regional life cycle assessment and economic analysis of camelina biodiesel production in the Pacific northwestern US. Journal of Cleaner Production. 2018;**172**:2389-2400

[7] Siddiquee MN, Rohani S. Lipid extraction and biodiesel production from municipal sewage sludge: A review. Renewable, and Sustainable Energy Reviews. 2011;**15**:1067-1072

[8] Kargbo DM et al. Biodiesel production from municipal sewage sludges: A review. Energy & Fuels. 2010;**24**(5):2791-2794

[9] Bharathiraja B, Yogendran D, Ranjith Kumar R, Chakravarthy M, Palani S. Biofuels from sewage sludge- a review. International Journal of ChemTech Research. 2014;**6**(9):4417-4327

[10] Thakur IS, Kumar M, Ghosh P, Khoslala K. Biodiesel production from municipal secondary sludge. Bioresource Technology. 2016;**216**:165-171

[11] Olkiewicz M, Fortuny A, Stüber F, Fabregat A, Font J, Bengoa C. Evaluation of different sludges from WWTP as a potential source for biodiesel production. Procedia Engineering. 2012;**42**:634-643

[12] Bharathi P, Pennarasi M. Production of lipids from municipal sewage sludge by two-stage extraction process. AJCHE. 2016;**16**(2):38-43

[13] Chaturvedi S, Dave PN, Shah NK. Applications of nano-catalyst in a new era'. Journal of Saudhi Chemical Society. 2012;**16**:307-325

[14] Jatav GK, De N. Application of nano-technology in soil plant system. An Arian Journal of Soil Sciences. 2013;**8**(2):176-184

[15] Ghormade V, Deshpande MV. Perspectives for nano-biotechnology enabled protection and nutrition of plants. Biotechnology Advances. 2011;**8**:35-45

[16] Singh MD, Chirag G, Prakash PO. Nano fertilizers is a new way to increase nutrients use efficiency in crop production. International Journal of Agricultural Sciences. 2017;**135**:78-98

[17] Duhan JS, Kumar R, Kumar N. Nanotechnology: The new perspective in precision agriculture. Biotechnology Reports. 2017;**15**:11-23

[18] Rahman KM, Melville L, Huq SI, Khoda SK. Understanding bioenergy production and optimization at the nanoscale – A review. Journal of Experimental Nanoscience. 2016;**11**(10):762-775

[19] Tiwari DK, Behari J, Sen P. Application of nanoparticles in waste water treatment. World Applied Sciences Journal. 2008;**3**(3):417-433

[20] Xiaolei Q, Alvarez PJJ, Li Q. Applications of nanotechnology in water and wastewater treatment. Water Research. 2013;**47**:3931-3946

[21] Endalew KE, Kiros Y, Zanzi R. Inorganic heterogeneous catalysts for biodiesel production from vegetable oils. Biomass and Bioenergy. 2011;**35**:3787-3809

[22] Nowatzki J, Andrew S, Dennis WP. Small- scale biodiesel production and use. NDSU Extension Service. 2007;**1344**:1-8

[23] Barabasand I, Todorut L-A. Predicting the temperature-dependent viscosity of biodiesel – Diesel – Bioethanol blends. Energy & Fuels. 2011;**25**(12):5767-5774

[24] Knothe G, Matheaus AC, Ryan TW III. Cetane numbers of branched and straight chain fatty esters determined in an ignition quality tester. Fuel. 2003;**82**:971-975

[25] Demirbas A. Biodiesel production from vegetable oils via catalytic and noncatalytic supercritical methanol transesterification methods. Progress in Energy and Combustion. 2005;**31**:466-487

[26] Dunn RO. Effect of antioxidants on the oxidative stability of methyl coyote (biodiesel). Fuel Processing Technology. 2005;**86**:1071-1085

[27] Lapuerta M, Rodrıguez-Fernandez J, Font de Mora E. Correlation for the estimation of the cetane number of biodiesel fuels and implications on the iodine number. Energy Policy. 2009;**37**:4337-4344

[28] Knothe G. "Designer" biodiesel: optimizing fatty ester composition to improve fuel properties. Energy & Fuels. 2008;**22**:1358-1313

[29] Allen CAW, Watts KC, Ackman RG, Pegg MJ. Predicting the viscosity of biodiesel fuels from their fatty acid ester composition. Fuel. 1999;**78**:1319-1326

[30] Pratas MJ et al. Biodiesel density: Experimental measurements and prediction models. Energy & Fuels. 2011;**25**:2333-2340

[31] Ashraful AM, Masjuki HH, Kalam MA, Fattah IMR, Imtenan S, Shahir SA, et al. Production and comparison of fuel properties, engine performance, and emission characteristics of biodiesel from various non-edible vegetable oils: A review. Energy Conversion and Management. 2014;**80**:202-208

[32] Rizwanul Fattah IM, Masjuki HH, Liaquat AM, Rahizar Ramli MA, Kalam VNR. Impact of various biodiesel fuels obtained from edible and non-edible oils on engine exhaust gas and noise emissions. Renewable and Sustainable Energy Reviews. 2013;**18**:552-567

[33] Palash SM, Kalam MA, Masjuki HH, Masum BM, Fattah IMR, Mofijur M. Impacts of biodiesel combustion on NOx emissions and their reduction approaches. Renewable and Sustainable Energy Reviews. 2013;**1**:473-490

[34] Ampah JD, Yusuf AA, Agyekum EB, Afrane S, Jin C, Liu H, et al. Progress

and recent trends in the application of nanoparticles as low carbon fuel additives—A state of the art review. Nanomaterials. 2022;**2**:1-59

[35] Gebremariam SN, Marchetti JM. Economics of biodiesel production: Review. Energy Conversion and Management. 2018;**168**:74-84

[36] Peters MS, Timmerhaus HD, West RE, Timmerhaus K, West R. Plant Design and Economics for Chemical Engineers. 4th ed. McGraw Hill International; 1991

[37] Gaurav A, Ng FT, Rempel GL. A new green process for biodiesel production from waste oils via catalytic distillation using a solid acid catalyst-modeling, economic and environmental analysis. Green Energy Environment. 2016;**1**:62-74

[38] Beckman C, Junyang J. People's Republic of China Biofuels Annual Report. 2009;**13040**:1-12

[39] Dufey A, Vermeulen S, Vorley B. Biofuels: Strategic choices for commodity-dependent developing countries. Common Fund for Commodities. 2007;**1**:72

[40] Bhattacharyya SC. The Economics of Renewable Energy Supply, Energy Economics: Concepts, Issues, Markets, and Governance. London, UK: SpringerVerlag; 2011. p. 249e71

[41] Mohammadshirazi A, Akram A, Rafiee S, Kalhor EB. Energy and cost analyses of biodiesel production from waste cooking oil. Renewable and Sustainable Energy Reviews. 2014;**33**:44-49

[42] Naveenkumar R, Baskar G. Optimization and techno-economic analysis of biodiesel production from Calophyllum inophyllum oil using heterogeneous catalyst. Bioresource Technology. 2020;**315**:123852

[43] Saratale GD, Saratale RG, Oh SE. Production and characterization of multiple cellulolytic enzymes by isolated Streptomyces sp. MDS. Biomass and Bioenergy. 2022;**5**:302-315

[44] Ricciardi R, Huskens J, Verboom W. Nanocatalysis in flow. Chemistry Sustainability Energy Materials Reviews. 2015;**8**:2586-2605

[45] Thangaraj B, Solomon PR, Muniyandi B, Ranganathan S, Lin L. Catalysis in biodiesel production – A review. Clean Energy. 2019;**3**:2-23

[46] Fattah IMR, Ong HC, Mahlia TMI, Mofijur M, Silitonga AS, Rahman SMA, et al. State of the art of catalysts for biodiesel production. Frontiers in Energy Research. 2020;**8**:101

[47] EIA. International Energy Outlook 2016, with Projections to 2040. Washington, DC: U.S. Energy Information Administration, Department of Energy; 2016

[48] Baskar G, Soumiya S. Production of biodiesel from castor oil using iron (II) doped zinc oxide nanocatalyst. Renewable Energy. 2016;**98**:101-107

Chapter 6

Bifunctional Heterogeneous Catalysts for Biodiesel Production Using Low-Cost Feedstocks: A Future Perspective

Welela Meka Kedir

Abstract

Biodiesel can be produced using domestic resources like straight vegetable oil, animal fats, and waste cooking oil. Its use, instead of conventional diesel, contributes to the reduction of CO_2 emissions. The production of biodiesel through transesterification (TE) reactions requires adequate catalysts to speed up the reactions. The classical methods of biodiesel production were conducted using homogeneous catalysts, which have drawbacks such as high flammability, toxicity, corrosion, byproducts like soap and glycerol, and a high wastewater output. Recently, various types of heterogeneous catalysts and continuous reactors have been invented for the production of biodiesel. As a result, the initial choice of catalysts is crucial. However, it is also affected by the amount of free fatty acids in a given sample of oil. In addition, most of the catalysts are not suitable for large-scale industrial applications due to their high cost. Bifunctional heterogeneous catalysts are widely applicable and have a rich history of facilitating energy-efficient, selective molecular transformations, and contributing to chemical manufacturing processes like biodiesel. This chapter underlines the use of bifunctional heterogeneous catalysts for biodiesel production using low-cost feedstock. Furthermore, it examines the sustainability of catalysts and low-cost feedstock for large-scale biodiesel production. Finally, the chapter indicates a further perspective of biodiesel as an alternative fuel using low-cost feedstock and recommends a sustainable bifunctional heterogeneous catalyst for biodiesel production.

Keywords: energy, biodiesel, bifunctional heterogeneous catalysts, transesterification, low-cost feedstock

1. Introduction

Because of the energy and global warming crisis, the development of renewable energy has been focused on worldwide [1]. Fossil fuel is the single largest energy source, representing 88% of all total world energy consumption [2]. The U.S. energy information administration, in its international energy outlook 2016 report, indicated that the world's total energy consumption is significantly increasing [3]. However, numerous studies showed that the combustion of non-renewable fossil fuels contributes approximately 52% of CO_2 emissions, which is the major source of greenhouse gases [4]. Nowadays, various renewable resources such as wind, geothermal, solar, wave energy, and biofuel are considered as alternative fossil fuels [5, 6]. Among these alternative fuels, biodiesel is being promoted as a supplementary fuel for diesel engines. The major benefits of biodiesel are its renewability, biodegradability, and ability to blend with other energy sources compared to other alternative fuels [7]. Even though biodiesel is a good alternative to petroleum diesel in various aspects, it is always jeopardized by the high cost of feedstock and the absence of economically and technically viable technology for its efficient production from feedstock [8, 9]. Biodiesel is also known as fatty acid methyl ester and is obtained by the transesterification reaction of methanol and vegetable oil in the presence of a suitable homogeneous or heterogeneous catalyst [10, 11]. Nowadays, various heterogeneous alkali catalysts such as zeolite, alkali earth metal oxides, KF/Al_2O_3, sodium aluminate etc. were developed for biodiesel production [9]. However, due to the expensive cost of catalyst synthesis, only selective heterogeneous catalysts were utilized in the industry [12].

Bifunctional catalysts facilitate the esterification of free fatty acids (FFAs) into alkyl esters alongside the transesterification (TE) reaction, which allows for the use of waste vegetable oils with high water and FFA contents for biodiesel production [13]. Under ideal reaction circumstances, acid-base bifunctional mixed-metal oxide catalysts produced biodiesel with a conversion rate of almost 100% [14]. For sustainable biomass upgrading, bifunctional heterogenous catalysts have the advantage of combining multiple catalytic processes in a single vessel. Additionally, a catalyst with an acid and base active phase will convert high FFA (>3%) feedstock in a single-step reaction through a simultaneous TE and esterification reaction process [15]. Mesoporous morphology dramatically enhanced the number of active sites, which helps to increase overall biodiesel production. Additionally, it is readily recyclable by straightforward washing and drying to remove adsorbed materials, sustaining activity for numerous cycles without showing any signs of metal leaching [16]. In order to employ waste vegetable oils with higher moisture and FFA levels for biodiesel generation, bifunctional solid catalysts enable the esterification of free fatty acids (FFA) into alkyl esters through the TE reaction [17]. Bifunctional heterogeneous catalysis is based on the fundamental idea that two different kinds of active sites cooperate to carry out a surface-catalyzed reaction [18]. Such two kinds of locations are frequently anticipated to catalyze several fundamental steps inside an overall reaction [19, 20]. Bifunctional materials could be recycled repeatedly, with only a slight deactivation that was attributed to the leaching of the various active metals during TE and poisoning by strongly adsorbed organics [8]. Before discussing the usage of comparable types of materials in biomass refining reactions, we highlight two important types of materials that are utilized as bifunctional catalysts in this section [21, 22]. Because they have enough acid sites on the surfaces with different strengths of Lewis acidity, Nafion-NR50, WO_3-ZrO_2, and SO_4-ZrO_2 are used for TE instead of other strong acid catalysts like hydrochloric, sulfuric, or phosphoric acid, hetero-poly acid inseminated on various supports (zirconia, silica, activated carbon, and alumina), and hydrochloric, sulfuric, or phosphoric acid [21]. Vegetable oil derived from edible plants, including palm, soybean, and sunflower oil are common feedstock in biodiesel production [23]. Currently, biodiesel production using edible vegetable oil as a raw material was the main cause of increased global food market prices [9]. Another issue related to edible oil bioenergy is the potential depletion of biomass resources as a result of intensive agricultural practices used in crop cultivation [4]. Numerous studies identified feedstock prices as the most important factor influencing the economic viability of the biodiesel market, accounting for up to 70–95% of total biodiesel production costs [15]. As a result, in order supply marketable biodiesel, the cost of the raw materials must be a key parameter [24]. The prominent alternate solution were using non-edible, low-cost feed stocks resource as a row material to investigate biodiesel [24–26]. Even though numerous study conducted on the production of biodiesel. The processing technologies are not yet commercially available, however, its predicted to enter the market within the next few years [27].

2. Biodiesel as alternative fuel

Biodiesel is extremely biodegradable and has a low toxicity level [23]. It emits nearly no aromatic chemicals or other chemical pollutants that are harmful to the environment [28]. When the entire life-cycle is evaluated (including cultivation, oil production, and oil conversion to biodiesel), it has a modest net contribution of carbon dioxide; and its production may be decentralized, it has tremendous potential for improving rural economies [29]. In comparison to diesel fuel, biodiesel emits no sulfur, less carbon monoxide, fewer particulates, less smoke and hydrocarbons, and more oxygen. More free oxygen results in complete combustion and lower emissions [30]. Currently, many countries use renewable energy like biodiesel (**Figure 1**).

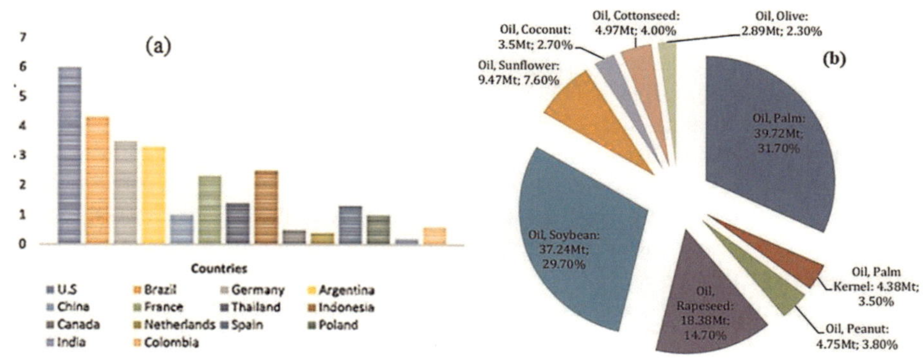

Figure 1.
Global biodiesel Trend (a) and worldwide use of vegetable oil (b) for biodiesel production [31, 32].

3. Production of biodiesel from different feedstocks

One of the advantages of producing biodiesel as an alternative fuel lies in its wide range of available feedstock [33]. The feedstock for biodiesel can be different from one country to another depending on their geographical locations and agricultural practices [30, 33]. The best feedstock must be chosen to guarantee minimal manufacturing costs. The ideal characteristics of a biodiesel feedstock include high oil content, ideal FFA composition, affordable agricultural resources, predictable growth and harvesting seasons, constant seed maturation rates, and a potential market for agricultural by-products [30, 34]. Currently, both edible and non-edible oils are used to produce biodiesel; however, using edible oils on a large scale poses a significant risk to the world [32, 35]. The use of biodiesel feedstock varies from country to country since it depends on the availability and the cost of biodiesel production were depicted in **Figure 2**.

Figure 2.
Production of biodiesel around the world from different raw materials [34].

4. Transesterification (TE)

A chemical process called TE turns TG and alcohol into alkyl esters and glycerol. It is a useful method for converting oil and fat feedstock, which chemically mimics petroleum diesel, into biodiesel. This method converts oils (TG) to low-viscosity alkyl esters, which are similar to diesel fuel [36]. This material can, therefore, be used in current petroleum-based diesel engines without needing to be modified because it has properties similar to those of petroleum-based diesel fuel. Reactants are frequently combined in TE, a reversible reaction when they are heated. But if a catalyst is added, the reaction will proceed more quickly [15]. The simplest chemical reaction for TE of TG is presented in **Figure 3**.

TE of oil and animal fats with a sufficient catalyst is a useful procedure for biodiesel production [37]. Several chemical catalysts are being utilized for TE of oil.

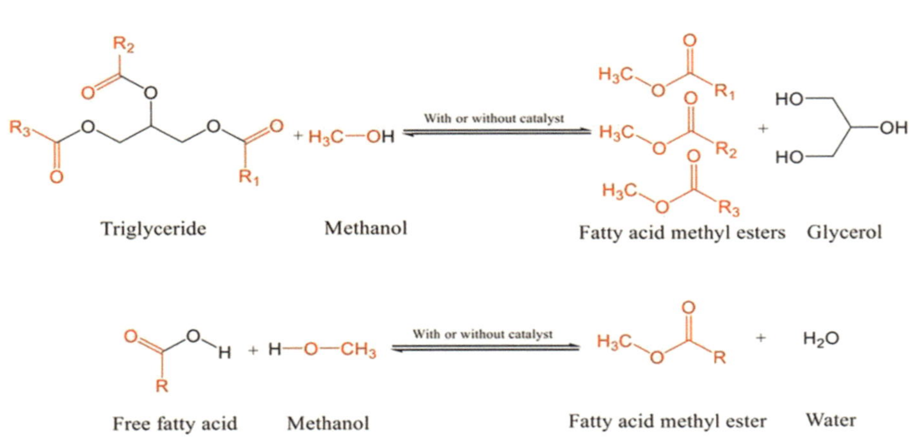

Figure 3.
Production of biodiesel from methanol and vegetable oil or animal fat.

However, these chemicals are expensive, scarce, poisonous, and ineffective. TE is typically catalyzed chemically, as in base catalyzed TE and acid-catalyzed TE, or via enzyme catalysts, as in lipase-catalyzed TEs [38]. When alcohol, often methanol, is used in non-catalyzed TE, there is no need for a catalyst because the alcohol is used in supercritical conditions, when the alcohol is at a temperature and pressure above its critical point and there is no separation between the liquid and gas phases [39]. In the supercritical state, the dielectric constant of alcohol is decreased so that the two-phase formation of vegetable oil/alcohol mixture is not encountered and only a single phase is found favoring the reaction [40]. Each TE process necessitates a unique feedstock. Some esterification procedures are more advantageous than others, at least in terms of manufacturing costs, waste creation, productivity, and so on [40, 41].

5. Catalyst for transesterification

The transesterification of oil can be catalytic, non-catalytic, or enzymatic. Catalytic transesterification of TG to fatty acid methyl ester was a major strategy for increasing biodiesel yield because the catalyst accelerates the rate at which the chemical reaction approaches equilibrium without becoming permanently involved [7]. Even when only a small amount of a catalyst is used, it affect the transesterification rate. The selectivity of catalysts, which are not consumed and many applications have been developed [42].

5.1 Classification of catalyst

Catalysts may be classified generally according to their physical state, their chemical nature, or the nature of the reaction that they catalyze (**Figure 4**). Catalysts used for biodiesel production are categorized into two types: heterogeneous catalysts and homogenous catalysts.

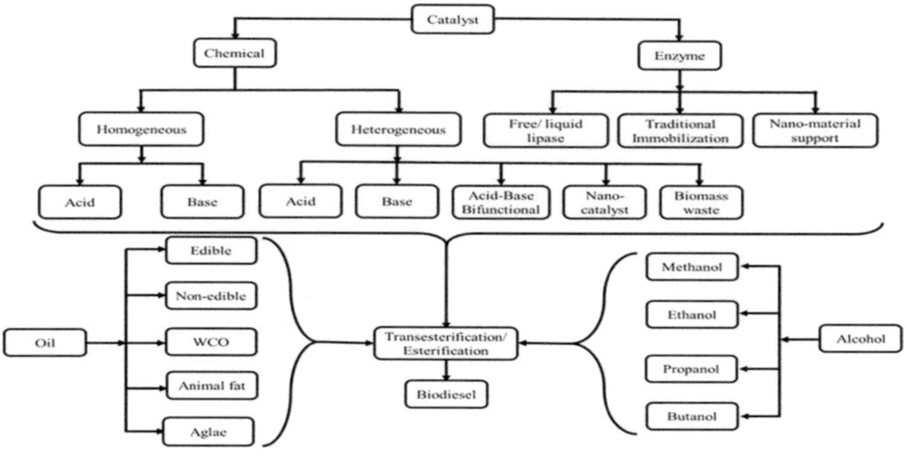

Figure 4.
Classification of catalysts, feedstock oils, and solvents used for biodiesel production via TE/esterification reactions [17].

5.1.1 Heterogeneous catalyst

In the heterogeneous TE reaction [8], a glyceride combines with an alcohol in the presence of a heterogeneous catalyst to produce fatty acid alkyl esters (biodiesel) and glycerol. The oxides of base supported on a large surface area, such as calcium oxide (CaO), magnesium oxide (MgO), and titanium dioxide, are commonly employed as heterogeneous catalysts in TE reactions [43, 44]. CaO is preferred as a catalyst because of its high activity, longer lifespan, and lack of consumption during the reaction [35]. Heterogeneous catalysts are understood to enhance the TE process by avoiding the extra processing costs associated with homogeneous catalysis and minimizing pollutant production (**Table 1**). Heterogeneous catalysts facilitate facile recovery, reusability, and a low-cost green process [15]. Efficient and low-cost heterogeneous catalysts help to reduce overall biodiesel production costs [25]. Heterogeneous catalysts are essential in difficult environments such as high temperatures and pressures. Such catalysts are easy to recover from the reaction mixture, can tolerate aqueous treatment stages, and can be modified to provide high activity, selectivity, and longer catalyst lifetimes. Several recent studies have focused on the technological and economic viability of producing biodiesel by heterogeneous acid-catalyzed TE [76]. As a result, the acidic catalytic reaction is very appealing for biodiesel production; however, acid catalysts exhibit lower catalytic performance in TE processes compared to basic catalysts, and heterogeneous solid catalysts were easily removed from the products in laboratory conditions. The water-washing and neutralizing processes were restricted [8].

5.1.2 Homogenous catalyst

Homogeneous chemical catalysts have good selectivity, high turnover frequency, high reaction rate, and easy activity adjustment [26]. Homogeneous chemical catalysts such as NaOH, CH_3ONa, and KOH were the most commonly used alkali catalysts (**Table 1**). Because of its high quality and low cost, homogeneous catalysts such as NaOH was used in transesterification; additionally, a small amount is required compared to KOH [42]. Currently, most of the heterogynous catalyst are not effective for the production of biodiesel which leads to hydrolysis or saponification of the fay acid methyl ester [7]. The resulting soap decreases the biodiesel yield and complexes the separation process. As a result, a two-step TE with acid first and alkali second was proposed [77]. The initial acid-based esterification efficiently reduces the oil's FFA content and prepares the oil for alkali catalysis.

5.2 Heterogeneous catalyst for biodiesel production

Many investigations have been conducted into the synthesis of heterogeneous catalysts in order to alleviate the issues associated with homogeneous catalysts in biodiesel synthesis. The literature has numerous reports on heterogeneous catalysts (acidic, basic, and enzymatic) for biodiesel synthesis. Alkali metal oxides and derivatives, as well as alkaline earth metal oxides [78], derived waste material-based heterogeneous catalysts [44], are examples of these ion exchange resins and sulfated oxides.

5.3 Heterogeneous catalyst from solid waste

Heterogynous catalysts synthesized from waste material play a crucial role to reduce organic pollutant elimination. Conversion of waste biomasses to catalysts helps

Optimization condition						
Catalyst	**MEOH: oil**	**Temperature (°C)**	**Catalyst (%wt.)**	**Time (h)**	**Conversion (%)**	**Reference**
CaO from eggshell	6:1	65	4	3	75.85	[25]
KOH load Snail shell	9:1	65	6	3.5	96	[37]
KF/CaO	8:1	65	5	2	95	[6]
Pyrolyzed rice husk	20:1	110	5	3	98.17	[45]
Li loaded eggshell	10:1		5	4	94	[46]
TiO_2–ZnO	6:1	60	200 mg	5	92.2	[36]
Scallop shell	6:1	65	5	-	86	[47]
KBr loaded eggshell	12:1	65	3		82.48	[48]
KI loaded oyster shell	6:1	60	3.5		85	[49]
SBA-15, AlSBA-15, with K_2CO_3, K_2SiO_3 and Kac	9:1	60	30	2.30	95	[50]
Snail shell modified with TiO_2–ZnO	6:1	65	3	3	90-95	[44]
Activated Carbon with K_2CO_3	15:1	65	5	5	85.1%	[51]
Citrus sinensis peel ash-coated magnetic nanoparticles	6:1	65	6	3	98	[52]
CaO/Fe_3O_4	15:1	65	1.3	2	95	[53]
Li/MgO, KOH/MgO	12:1	60	9	2	93.9	[54]
Mg-Al hydrotalcite	4:1	45	1	1.5	95.2	[55]
Dolomite, CaMgO and CaZnO	6:1	67.5	3	3	91.78	[56]
KF/Ca-Al hydrotalcite	12:1 5	65	5	5	97.98	[57]
CaO/mesoporous silica	16:1	60	5	8	95.2	[58]
KNO_3/CaO-MgO	6:1	52.5	0.9	3	78	[59]
CaO/Al_2O_3	12:1	65	100.54	5	98.64	[60]
Nano- KF/Al_2O_3	8:1	65	5	2	98.8	[61]
Biochar CaO/Al_2O_3	18:1	65	3	3.25	98.3	[62]
CaO/dolomite	6:1	65	5	3	90	[63]
MgO/$MgAl_2O_4$	12:1	110	3	3	95	[64]
$S_2O_8^{2-}$/ZrO_2	20:1	110	3	4	100	[65]
Sulfated zirconia	20:1	150	3	6	100	[66]
SO3H-ZnAl2O4	5:1	200	1	3	74	[67]
HZSM-5	9:1	120	1	1	94.7	[68]
Carbon cryogel	20:1	65	5	5	91.3	[69]
TiSBA-15-Me- $PrSO_3H$	90:1	65	5	9	71	[70]
Cs-Na_2ZrO_3	30:1	65	1	0.25	98.8	[71]
40K/PA-550	12:1	65	5	3.5	97	[31]

Optimization condition						
Catalyst	**MEOH: oil**	**Temperature (°C)**	**Catalyst (%wt.)**	**Time (h)**	**Conversion (%)**	**Reference**
35KOH/ZSM5	12:1	60	18	8	93.8	[72]
Sodium silicate	12:1	65	2.5	0.5	97	[73]
$MgO/MgAl_2O_4$	12:1	110	3	3	-	[74]
La/Mn oxide	12:1	180	3	1.5	-	[75]

Table 1.
Comparison of various solid catalysts for transesterification reaction.

to improve environmental sustainability and renewable energy production such as biodiesel. CaO derived from seashells, chicken egg shells, and crab shells has been identified as an effective heterogeneous catalyst for biodiesel production (**Tables 1** and **2**). Those certain shells were reached in calcium carbonate (95%), with the rest being organic materials and other substances like $MgCO_3$, phosphate, and trace metals [81]. Activated catalysts derived from calcined mud crab shells and waste cockle shells reacted with a 3 h reaction rate using palm oil, yielding 98.2% and 99% biodiesel, respectively [37]. Moreover, 98.5% FAME was derived in the TE reactions of palm oil using river snail shells as catalysts (>800°C). With 90% producing biodiesel, CaO leaching is also revealed to be the primary cause of catalytic activity. The CaO impregnated in deionized KF produces 85% biodiesel yield from soybean oil in 4 h with a catalyst concentration of 3.5% wt., with 1:6 oil-to-methanol molar ratio and a reaction temperature of 60°C [82].

5.4 Acid-catalyzed transesterification

Heterogeneous acid catalysts are less corrosive and harmful than homogeneous acid catalysts and cause fewer environmental problems [83]. These catalysts have a wide range of acidic sites (**Figure 5**) with varying degrees of Brønsted or Lewis acidity. While these catalysts show promising performance under mild reaction conditions, they react much more slowly than solid-base catalysts [19]. Furthermore, this type of catalyst requires a large catalytic loading, high temperature, and a long reaction time.

5.5 Base-catalyzed transesterification

The most widely utilized approach in the industry is the employment of alkaline catalysts in the TE of waste cooking oil. Because of their lower cost, metallic hydroxides are commonly utilized as catalysts (**Figure 6**). However, they have less activity than alkoxide [35]. According to reports, the pace of a base-catalyzed reaction is 4000 times faster than that of an acid-catalyzed reaction [83]. The most widely employed catalysts are potassium hydroxide (KOH) and sodium hydroxide (NaOH), both of which are highly sensitive to the quality of the reaction being impacted by the presence of water and free fatty acids [9]. In an alkaline environment, the presence of water may undergo saponification rather than esterification. Furthermore, the free fatty acids can react with the alkaline catalyst to produce soap and water [40].

Catalyst types		Examples	Advantages	Disadvantages
Homogeneous	Alkali	NaOH, KOH	• High reactivity • High reaction rate • Cheaper cost • Medium temperature condition • Greater yield • Cheaper operating conditions to the TE for industrial scale	• Inappropriate for greater FFA in feedstocks and moisture • Deactivates in the presence of wastewater, necessitates a large amount of reaction, saponification happens as a by-product, is non-recyclable • Expensive, difficult to separate from the finished product
	Acid	H_2SO_4, HCl, HF	• Non-reactive to moisture and FFA content in oil • Catalyzed simultaneous • Avoids saponification • Greater esterification reaction performance	• Slow rate of reaction • Corrosive and large reaction time • Higher reaction temperature and pressure, high alcohol/oil requirement • Weak catalytic activity, catalyst is difficult to recycle, very slow for TE reaction • Difficulty separating from the product
Heterogeneous	Alkali	CaO, SrO, MgO, mixed oxide, and hydrotalcite	• Non-corrosive • Environmental eco-friendly • Easy separate • Fewer disposal problems • Recyclable • Higher selectivity • Longer catalyst life • Simple separation • Reactivity and reusability • long lifetime. • High performance for transesterification reaction under normal conditions.	• Slow reaction rate compared to homogeneous one feedstock (<1 wt. %) • Low FFA requirement • Saponification as a side reaction • Highly sensitive to moisture and FFA • High volume of wastewater • Soap formation • Diffusion limitations route Complex and expensive synthesis. • High moisture attraction during storage • Inappropriate for Feedstocks containing high free fatty acids. • Not tolerant of free fatty acid and water • High leaching • Not economical.
	Acid	ZrO, TiO, ZnO, ion exchange resin, sulfonic modified Meso structured silica.	• Insensitive to FFA and moisture content • Simultaneously catalyze esterification and transesterification • Recyclable • Easy to separation • Long lifetime • Compatible with esterification reaction • Environmental eco-friendly	• High reaction time • Greater reaction temperature and pressure • High alcohol/oil needed • Lower catalytic activity • Low acidic site and microporosity • Complex and non-economical synthesis route • Expensive cost of catalyst • Minimum activity and reaction rate • Weak catalytic performance • High temperatures and long reaction times required

Table 2.
Benefits and drawbacks of various types of catalysts used for biodiesel production [79, 80].

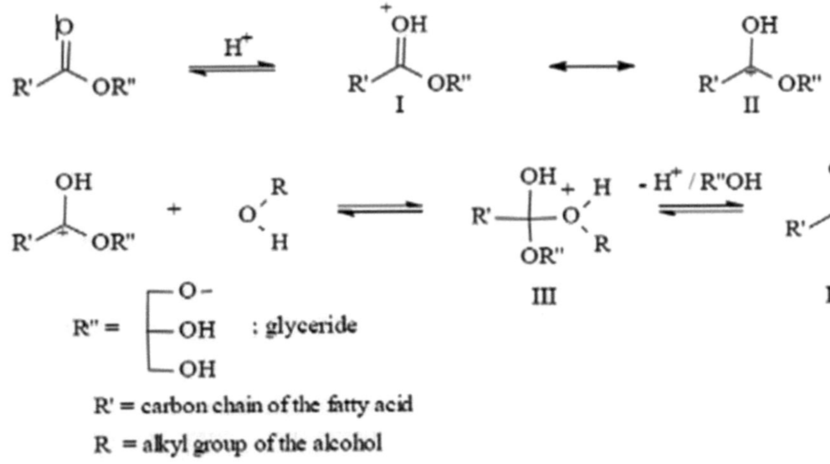

Figure 5.
The mechanism of the acid-catalyzed TE of vegetable oils [84, 85].

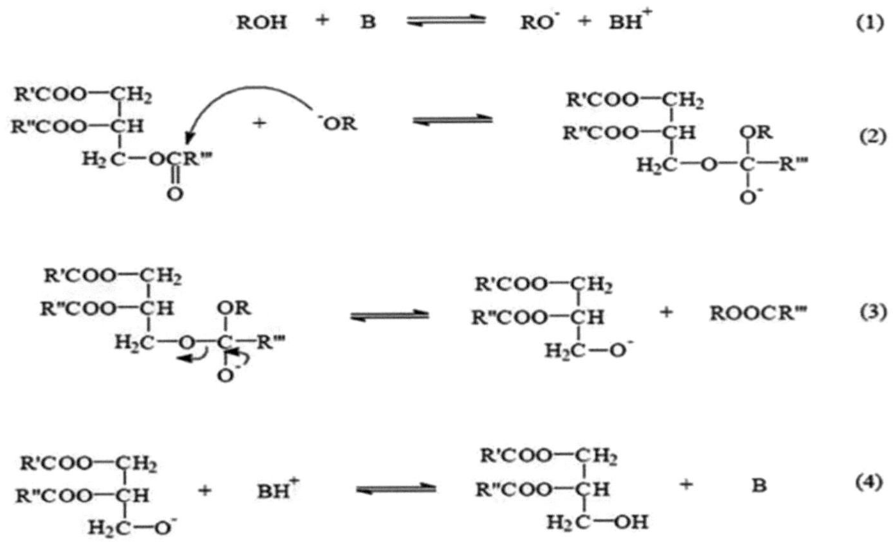

Figure 6.
The mechanism of the base-catalyzed TE of vegetable oils [84, 85].

Saponification not only depletes the catalyst but also results in the production of emulsions that impede biodiesel separation, recovery, and purification.

6. Bifunctional heterogeneous catalyst

Bifunctional heterogeneous catalysts are highly preferred by the chemical industry for both financial and environmental reasons, provided that they are available in the required quantity at a reasonable price and exhibit notable operational advantages over the corresponding homogeneous system [18]. Only a few examples of the use of heterogeneous multifunctional catalysts have been documented, and the majority of them, especially those containing transition metals, have issues with product contamination and limited reusability [20]. Additionally, they frequently require excessive amounts of expensive reagents or other additives, making them unsuitable for the mass production of extensive organic compounds [86]. Bifunctional catalysts that contain both supported acid and metal sites may encourage the transesterification reaction (**Table 1**). There are two types of bifunctional heterogeneous catalysts. i. Bimetallic: One of the most popular methods of regulating the catalytic performance of supported metal catalysts is to control the composition of bimetallic catalysts [21]. The intention of adding one metal to another is frequently to change the activity of the original metal rather than to establish a new catalytic function. The assumption that the modifying metal is absent from the surface layer limits its ability to directly produce a bifunctional effect in many of the situations where it is dominant [87]. Studies of these bifunctional (or "cooperative") catalysts have attracted a lot of attention in the scientific community [11]. There was a significant amount of interest in biodiesel production based on numerous reports in the literature on bifunctional heterogeneous catalysts. However, efficient and effective resources should be used to synthesize those catalysts. Furthermore, for large-scale industrial applications, optimization parameters such as reaction rate and effective transesterification should be required. In addition to focusing on efficient oil-to-biodiesel conversion, waste material utilization as a bifunctional heterogynous catalyst should be required to address the current associated problem of environmental pollution. As a result, there has been a lot of research done on bifunctional heterogeneous catalysts. In general, the use of bifunctional (or multifunctional) catalysts is a big step in the right direction for catalysis, especially for reactions that necessitate various kinds of intermediates as part of an overall reaction (**Table 1**).

7. Current status and future research direction

The main source of greenhouse gases is CO_2 emissions, which account for around 52% of emissions. Therefore, in order to reduce the greenhouse effect, alternative energy sources or technology must be developed. Alternative fossil fuels include a variety of renewable resources like wind, geothermal, solar, wave energy, and biofuel. Low efficiency and a high initial maintenance cost are a few downsides. Given the different difficulties of using renewable energy, biodiesel is a possible replacement for diesel made from petroleum. Due to its superior performance, biodegradability, non-toxic nature, lack of hazards, carbon neutrality, low pollution, environmental friendliness, low flammability, superior transportation, longer storage time, and low CO production, biodiesel is significant. On the other hand, it is crucial to produce biodiesel from non-edible sources as well as from edible sources. For the manufacturing of biodiesel, it is best practice to transesterify vegetable oil and animal fats using the appropriate catalyst. Currently, the TE of oil uses numerous chemical catalysts. However, those substances are expensive, scarce, hazardous, and ineffective. To address the current difficulties, it is therefore very important to produce alternative energy sources (biodiesel) from deep-frying oil and heterogeneous catalysts from solid waste. The bifunctional heterogynous catalyst has been the subject of numerous investigations and optimizations. To create an effective and efficient biodiesel yield, the bifunctional heterogynous catalyst must still be produced and studied. Additionally, intensification reactors will call for kinetics studies, optimization parameters, and cost-effective TE in the presence of a bifunctional heterogynous catalyst, which may result in the creation of a workable replacement process for current industrial units. Furthermore, for efficient biodiesel synthesis and the minimization of harmful environmental consequences connected with waste materials, bifunctional heterogynous catalysts made from a variety of environmentally favorable materials would be needed. Therefore, more research should be conducted to produce biodiesel from low-cost feedstock using effective bifunctional heterogeneous catalysts. This requires particularly extensive research to demonstrate their potential for use in the synthesis of industrial biodiesel.

8. Conclusion

Currently, there are numerous types of catalysts that have been investigated for transesterification. However, most of them are expansive, ineffective, and produce a lower yield of biodiesel. To overcome these problems, a bifunctional heterogynous catalyst was a viable alternative for producing high biodiesel yields. As a result, numerous studies on bifunctional heterogeneous catalysts should be conducted in order to improve biodiesel yields and reduce environmental pollution. A bifunctional heterogeneous catalyst can simultaneously esterify free fatty acids and transesterify triglycerides in oil. Using edible oil as a feedstock is not economical and is a major concern for biodiesel production. Therefore, research should be conducted on biomass waste as an alternative feedstock for the production of biodiesel. The chapter summarizes recent research advances in the development of bifunctional heterogynous catalysts for cost-effective biodiesel production. To prepare a better yield of biodiesel, future researchers should be focused on low-grade feedstock and bifunctional heterogeneous catalysts as an effective new route to replacing diesel fuel as a viable energy source in the near future.

Author details

Welela Meka Kedir
Department of Chemistry, College of Natural and Computational Science,
Mattu University, Mattu, Ethiopia

*Address all correspondence to: welelameka008@gmail.com

References

[1] Faruque MO, Razzak SA, Hossain MM. Application of heterogeneous catalysts for biodiesel production from microalgal oil: A review. Catalysts. 2020;**10**(9):1-25

[2] Kalligeros S, Zannikos F, Stournas S, Lois E, Anastopoulos G, Teas C, et al. An investigation of using biodiesel/marine diesel blends on the performance of a stationary diesel engine. Biomass and Bioenergy. 2003;**24**:141-149

[3] Jamil F, Al-haj L, Al-muhtaseb AH, Al-hinai MA, Baawain M. Current scenario of catalysts for biodiesel production: A critical review. Review Chemical Engineering. 2017;**2017**:3-31

[4] El Sherbiny SA, Refaat AA, El Sheltawy ST. Production of biodiesel using the microwave technique. Journal of Advanced Research. 2010;**1**(4): 309-314

[5] Lani NS, Ngadi N, Taib MR. Parametric study on the transesterification reaction by using CaO/silica catalyst. Chemical Engineering Transactions. 2017;**56**: 601-606

[6] Thi T, Myat M. Synthesis and characterization of CaO and KF Doped CaO (KF/CaO) derived from chicken eggshell waste as heterogeneous catalyst in biodiesel production. American Science Research Journal of Engineering and Technology Science. 2017;**38**(2): 134-151

[7] Thangaraj B, Solomon PR, Muniyandi B, Ranganathan S, Lin L. Catalysis in biodiesel production—A review. Clean Energy. 2019;**3**(1):2-23

[8] Lee AF, Bennett JA, Manayil JC, Wilson K. Heterogeneous catalysis for sustainable biodiesel production via esterification and transesterification. Chemical Society Reviews. 2014;**43**(22): 7887-7916

[9] Agarwal M. Biodiesel production from a mixture of vegetable oils using marble slurry derived heterogeneous catalyst. Current Trends in Biomedical Engineering Bioscience. 2017;**5**(1):1-5

[10] Li Y, Jiang Y. Preparation of a Palygorskite supported KF/CaO catalyst and its application for biodiesel production via transesterification. RSC Advances. 2018;**8**(29):16013-16018

[11] Refaat AA. Biodiesel production using solid metal oxide catalysts. International Journal of Environmental Science and Technology. 2011;**8**(1): 203-221

[12] Birla A, Singh B, Upadhyay SN, Sharma YC. Kinetics studies of synthesis of biodiesel from waste frying oil using a heterogeneous catalyst derived from snail shell. Bioresource Technology. 2012;**106**:95-100

[13] Sivasamy A, Cheah KY, Fornasiero P, Kemausuor F, Zinoviev S, Miertus S. Catalytic applications in the production of biodiesel from vegetable oils. ChemSusChem. 2009;**2**(4):278-300

[14] Astar I, Usman T, Wahyuni N, Alimuddin AH. Bifunctional heterogeneous catalysts from oil palm empty fruit bunches ash and alum for biodiesel synthesis simultaneously. AIP Conference Proceedings. 2017;**1823**(03): 1-6

[15] Mansir N, Hwa Teo S, Lokman Ibrahim M, Yun Hin TY. Synthesis and application of waste egg shell derived CaO supported W-Mo mixed oxide

catalysts for FAME production from waste cooking oil: Effect of stoichiometry. Energy Conversion and Management. 2017;**151**(08): 216-226

[16] Jayakumar M, Karmegam N, Gundupalli MP, Bizuneh Gebeyehu K, Tessema Asfaw B, Chang SW, et al. Heterogeneous base catalysts: Synthesis and application for biodiesel production—A review. Bioresource Technology. 2021;**3**:331

[17] Mandari V, Devarai SK. Biodiesel production using homogeneous, heterogeneous, and enzyme catalysts via transesterification and esterification reactions: A critical review. Bioenergy Research. 2021;**1**(2):2-27

[18] Dembo MD. Synthesis of a bifunctional heterogeneous catalyst for the conversion of lignocellulosic biomass into precursors for alternative fuels and fine chemicals. 2013:1-67

[19] Wang A, Quan W, Zhang H, Li H. Heterogeneous ZnO-containing catalysts for efficient biodiesel production. RSC Advances. 2021:20465-20478

[20] Barbaro P, Liguori F, Linares N, Marrodan CM. Heterogeneous bifunctional metal/acid catalysts for selective chemical processes. European Journal of Inorganic Chemistry. 2012;**24**: 3807-3823

[21] Robinson AM, Hensley JE, Will Medlin J. Bifunctional catalysts for upgrading of biomass-derived oxygenates: A review. ACS Catalysis. 2016;**6**(8):5026-5043

[22] Elias S, Rabiu AM, Okeleye BI, Okudoh V, Oyekola O. Bifunctional heterogeneous catalyst for biodiesel production from waste vegetable oil. Applied Sciences. 2020;**10**(9)

[23] Atabani AE, Silitonga AS, Badruddin IA, Mahlia TMI, Masjuki HH, Mekhilef S. A comprehensive review on biodiesel as an alternative energy resource and its characteristics. Renewable and Sustainable Energy Reviews. 2012;**16**(4):2070-2093

[24] Azócar L, Ciudad G, Heipieper HJ, Navia R. Biotechnological processes for biodiesel production using alternative oils. Applied Microbiology and Biotechnology. 2010:621-636

[25] Ngadi N, Sulaiman S, Rahman RA, Lani NS. Production of biodiesel from palm oil using cockle shell waste as heterogeneous catalyst. Journal of Teknology. 2017;**79**(5):183-188

[26] Gnanaprakasam A, Sivakumar VM, Surendhar A, Thirumarimurugan M, Kannadasan T. Recent strategy of biodiesel production from waste cooking oil and process influencing parameters: A review. Journal of Energy. 2013;**2013**: 1-10

[27] Huang D, Zhou H, Lin L. Biodiesel: An alternative to conventional fuel. Energy Procedia. 2012;**16**(PART C): 1874-1885

[28] Ogunkunle O, Ahmed NA. Overview of biodiesel combustion in mitigating the adverse impacts of engine emissions on the sustainable human–environment scenario. Sustainability. 2021;**13**(10)

[29] Fazal MA, Haseeb ASMA, Masjuki HH. Biodiesel feasibility study: An evaluation of material compatibility; performance; emission and engine durability. Renewable and Sustainable Energy Reviews. 2011;**15**(2):1314-1324

[30] Hamzah NHC, Khairuddin N, Siddique BM, Hassan MA. Potential of Jatropha Curcas L. as biodiesel feedstock

in malaysia: A concise review. PRO. 2020;**8**(7):1-11

[31] Shan R, Shi J, Yan B, Chen G, Yao J, Liu C. Transesterification of palm oil to fatty acids methyl ester using K_2CO_3/ palygorskite catalyst. Energy Conversion and Management. 2016;**116**: 142-149

[32] Khan HM, Ali CH, Iqbal T, Yasin S, Sulaiman M, Mahmood H, et al. Current scenario and potential of biodiesel production from waste cooking oil in Pakistan: An overview. Chinese Journal of Chemical Engineering. 2019;**27**(10): 2238-2250

[33] Yahya NY, Ngadi N. Synthesis of mesoporous calcium titanate catalyst for transesterification of used cooking oil : A review of the synthesized potential. International Journal of Advanced and Applied Sciences. 2020;**7**(1):20-41

[34] Gardy J, Rehan M, Hassanpour A, Lai X, Nizami AS. Advances in nano-catalysts based biodiesel production from non-food feedstocks. Journal of Environmental Management. 2019;**2019**: 249

[35] Lee HV, Juan JC, Taufiq-Yap YH, Kong PS, Rahman NA. Advancement in heterogeneous base catalyzed technology: An efficient production of biodiesel fuels. Journal of Renewable Sustainable Energy. 2015;**2015**:7

[36] Madhuvilakku R, Piraman S. Biodiesel synthesis by TiO2-ZnO mixed oxide nanocatalyst catalyzed palm oil transesterification process. Bioresource Technology. 2013;**150**:55-59

[37] Gupta J, Agarwal M. Preparation and characterization of highly active solid base catalyst from snail shell for biodiesel production. Biofuels. 2019; **10**(3):315-324

[38] Mandolesi De Araújo CD, De Andrade CC, De Souza E, Silva E, Dupas FA. Biodiesel production from used cooking oil: A review. Renewable and Sustainable Energy Reviews. 2013; **27**:445-452

[39] Marulanda VF. Biodiesel production by supercritical methanol transesterification: Process simulation and potential environmental impact assessment. Journal of Cleaner Production. 2012;**33**:109-116

[40] Kusdiana D, Saka S. Effects of water on biodiesel fuel production by supercritical methanol treatment. Bioresource Technology. 2004;**91**(3): 289-295. DOI: 10.1016/S0960-8524(03) 00201-3

[41] Ismail SAEA, Ali RFM. Physico-chemical properties of biodiesel manufactured from waste frying oil using domestic adsorbents. Science and Technology of Advanced Materials. 2015;**2015**:16

[42] Endalew AK, Kiros Y, Zanzi R. Inorganic heterogeneous catalysts for biodiesel production from vegetable oils. Biomass and Bioenergy. 2011;**35**(9): 3787-3809

[43] Changmai B, Vanlalveni C, Ingle P. Widely used catalysts in biodiesel production. RSC Advances. 2020: 41625-41679

[44] Meka Kedir W, Girma Asere T. Biodiesel production from waste frying oil using catalysts derived from waste materials. JOTCSA. 2022;**9**(3):939-952

[45] Li M, Zheng Y, Chen Y, Zhu X. Biodiesel production from waste cooking oil using a heterogeneous catalyst from pyrolyzed rice husk. Bioresource Technology. 2014;**154**:345-348

[46] Boro J, Konwar LJ, Deka D. Transesterification of non edible feedstock with lithium incorporated egg shell derived CaO for biodiesel production. Fuel Processing Technology. 2014;**122**:72-78

[47] Sirisomboonchai S, Abuduwayiti M, Guan G, Samart C, Abliz S, Hao X, et al. Biodiesel production from waste cooking oil using calcined scallop shell as catalyst. Energy Conversion and Management. 2015;**95**:242-247

[48] Mahesh SE, Ramanathan A, Begum KMMS, Narayanan A. Biodiesel production from waste cooking oil using KBr impregnated CaO as catalyst. Energy Conversion and Management. 2015;**91**:442-450

[49] Jamil F, Al-Haj L, Al-Muhtaseb AH, Al-Hinai MA, Baawain M, Rashid U, et al. Current scenario of catalysts for biodiesel production: A critical review. Reviews in Chemical Engineering. 2018; **34**(2):267-297

[50] Wu H, Zhang J, Liu Y, Zheng J, Wei Q. Biodiesel production from jatropha oil using mesoporous molecular sieves supporting K_2SiO_3 as catalysts for transesterification. Fuel Processing Technology. 2014;**119**:114-120

[51] Fan M, Zhang P. Activated carbon supported K_2CO_3 catalysts for transesterification of dimethyl carbonate with propyl alcohol. Energy and Fuels. 2007;**21**(2):633-635

[52] Changmai B, Rano R, Vanlalveni C, Rokhum L. A novel citrus sinensis peel ash coated magnetic nanoparticles as an easily recoverable solid catalyst for biodiesel production. Fuel. 2021;**286** (P2):119-447

[53] Ali MA, Al-Hydary IA, Al-Hattab TA. Nano-magnetic catalyst CaO-Fe_3O_4 for biodiesel production from date palm seed oil. Bulletin Chemical Reactive Engineering Catalaysis. 2017;**12**(3): 460-468

[54] Rafati A, Tahvildari K, Nozari M. Production of biodiesel by electrolysis method from waste cooking oil using heterogeneous MgO-NaOH nano catalyst. Energy sources: Part A Recover. Utilities Environmental Efficient. 2019; **41**(9):1062-1074

[55] Di Serio M, Mallardo S, Carotenuto G, Tesser R, Santacesaria E. Mg/Al hydrotalcite catalyst for biodiesel production in continuous packed bed reactors. Catalysis Today. 2012;**195**(1): 54-58

[56] Ajala EO, Ajala MA, Odetoye TE, Okunlola AT. Synthesis of solid catalyst from dolomite for biodiesel production using palm Kernel oil in an optimization process by definitive screening design. Brazilian Journal of Chemical Engineering. 2019;**36**(2):979-994

[57] Gao L, Teng G, Xiao G, Wei R. Biodiesel from palm oil via loading KF/Ca-Al hydrotalcite catalyst. Biomass and Bioenergy. 2010;**34**(9): 1283-1288

[58] Albuquerque MCG, Jiménez-Urbistondo I, Santamaría-González J, Mérida-Robles JM, Moreno-Tost R, Rodríguez-Castellón E, et al. CaO supported on mesoporous silicas as basic catalysts for transesterification reactions. Applied Catalysis A: General. 2008;**334** (1–2):35-43

[59] Ibrahim Y, Idris J, Muhammed Z, Muhammad A. Synthesis, characterization and formation of KNO3 treated CaO-MgO composite catalysts for biodiesel production using neem seed oil. Journal of Material Environmental Science. 2020;**11**(5):781-787

[60] Zabeti M, Daud WMAW, Aroua MK. Optimization of the activity of CaO/Al2O3 catalyst for biodiesel production using response surface methodology. Applied Catalysis A: General. 2009;**366**(1):154-159

[61] Tang Y, Ren H, Chang F, Gu X, Zhang J. Nano KF/Al2O3 particles as an efficient catalyst for no-glycerol biodiesel production by coupling transesterification. RSC Advances. 2017; 7(10):5694-5700

[62] Konwar LJ. Biochar supported Cao as heterogeneous catalyst for biodiesel production. International Journal of Innovation Research Development. 2012;**1**(7):186-195

[63] Sutrisno B, Nafiah AD, Fauziah IS, Kurniawan W, Hinode H, Hidayat A. Cao/natural dolomite as a heterogeneous catalyst for biodiesel production. Materials Science Forum. 2020;**991**: 117-122

[64] Rahmani Vahid B, Haghighi M. Biodiesel production from sunflower oil over $MgO/MgAl_2O_4$ nanocatalyst: Effect of fuel type on catalyst nanostructure and performance. Energy Conversion and Management. 2017;**134**:290-300

[65] Wang H, Wei G, Liu F, Banerjee G, Joshi M. Characterization of Two Homogalacturonan Pectins with Immunomodulatory Activity from Green Tea. 2014;**2014**:9963-9978

[66] Shi G, Yu F, Wang Y, Pan D, Wang H, Li R. A novel one-pot synthesis of tetragonal sulfated zirconia catalyst with high activity for biodiesel production from the transesterification of soybean oil. Renewable Energy. 2016; **92**:22-29

[67] Adeniyi P, Muhammad Y, Yakub I, Abdalla Y, Mohd W, Wan A. Article in press synthesis and application of hierarchical mesoporous HZSM-5 for biodiesel production from Shea butter. Journal of the Taiwan Institute of Chemical Engineers. 2015:1-8

[68] Soltani S, Rashid U, Yunus R, Taufiq-Yap YH. Biodiesel production in the presence of sulfonated mesoporous $ZnAl_2O_4$ catalyst via esterification of palm fatty acid distillate (PFAD). Fuel. 2016;**178**(03):253-262

[69] Zainol MM, Amin NAS, Asmadi M. Synthesis and characterization of carbon cryogel microspheres from lignin-furfural mixtures for biodiesel production. Bioresource Technology. 2015;**190**:44-50

[70] Léon CIS, Song D, Su F, An S, Liu H, Gao J, et al. Propylsulfonic acid and methyl bifunctionalized TiSBA-15 silica as an efficient heterogeneous acid catalyst for esterification and transesterification. Microporous and Mesoporous Materials. 2015;**204**(C): 218-225

[71] Torres-Rodríguez DA, Romero-Ibarra IC, Ibarra IA, Pfeiffer H. Biodiesel production from Soybean and Jatropha oils using cesium impregnated sodium zirconate as a heterogeneous base catalyst. Renewable Energy. 2016;**93**: 323-331

[72] Saba T, Estephane J, El Khoury B, El Khoury M, Khazma M, El Zakhem H, et al. Biodiesel production from refined sunflower vegetable oil over KOH/ZSM_5 catalysts. Renewable Energy. 2016;**90**: 301-306

[73] Roschat W, Siritanon T, Yoosuk B, Promarak V. Rice husk-derived sodium silicate as a highly efficient and low-cost basic heterogeneous catalyst for biodiesel production. Energy Conversion and Management. 2016;**119**:453-462

[74] Rahmani Vahid B, Haghighi M. Urea-nitrate combustion synthesis of $MgO/MgAl_2O_4$ nanocatalyst used in biodiesel production from sunflower oil: Influence of fuel ratio on catalytic properties and performance. Energy Conversion and Management. 2016;**126**: 362-372

[75] Nasreen S, Liu H, Skala D, Waseem A, Wan L. Preparation of biodiesel from soybean oil using La/Mn oxide catalyst. Fuel Processing Technology. 2015;**131**(03):290-296

[76] Istadi I, Mabruro U, Kalimantini BA, Buchori L, Anggoro DD. Reusability and stability tests of calcium oxide based catalyst (K_2O/CaO-ZnO) for transesterification of soybean oil to biodiesel. Bulletin Chemical Reactive Engineering Catalyst. 2016;**11**(1):34-39

[77] Demirbas A, Bafail A, Ahmad W, Sheikh M. Biodiesel production from non-edible plant oils. Energy Exploration & Exploitation. 2016;**34**(2):290-318

[78] Vasi K, Podrepšek GH, Knez Ž, Leitgeb M. Biodiesel production using solid acid catalysts. Catalysts. 2020;**10**: 237

[79] Nomanbhay S, Ong MY. A review of microwave-assisted reactions for biodiesel production. Bioengineering. 2017;**4**(2)

[80] Hidalgo P, Toro C, Ciudad G, Navia R. Advances in direct transesterification of microalgal biomass for biodiesel production. Reviews in Environmental Science and Biotechnology. 2013;**12**(2): 179-199

[81] Suryaputra W, Winata I, Indraswati N, Ismadji S. Waste Capiz (Amusium Cristatum) shell as a new heterogeneous catalyst for biodiesel production. Renewable Energy. 2013;**50**:795-799

[82] Owusu PA, Asumadu-Sarkodie S. A review of renewable energy sources, sustainability issues and climate change mitigation. Cogent Engineering. 2016;**3** (1):1-14

[83] Fattah IMR, Ong HC, Mahlia TMI, Mofijur M, Silitonga AS. State of the art of catalysts for biodiesel production. Frontier in Energy Reassures. 2020;**8**(7): 1-17

[84] Montero G, Stoytcheva M, Garcia C, Coronado M, Toscano L, Campbell H, et al. Current status of biodiesel production in Baja California, Mexico. Biodiesel-Quality Emission By-Products. 2014; **2011**:10

[85] Dhanamurugan A, Subramanian R. Performance and emission characteristics of a diesel engine with various injection pressures using bael biodiesel. Applied Mechanics and Materials. 2014;**592–594**(1):1714-1718

[86] Maina L. Investigation of potential of a bifunctional catalyst for the simultaneous esterification and transesterification of high free fatty acid feedstock. Waste and Biomass Valorization. 2021;**2021**:1-86

[87] Di Serio M, Tesser R, Pengmei L, Dipartimento ES. Biochar-derived heterogeneous catalysts for biodiesel production. Environmental Chemistry Letters. 2019;**17**(4):1447-1469

Section 4

Biodiesel Postprocess

Chapter 7

Biodiesel Refining and Processing Strategies

Timothy J. Tse, Li Zhou, Farley Chicilo, Venkatesh Meda and Martin J.T. Reaney

Abstract

Biodiesel fuel is produced from triglyceride fats, and oils obtained from plant and animal sources. Typically, triglycerides are first transesterified to produce fatty acid alkyl esters (FAAE) and then refined. Traditional FAAE refining strategies are often energy-intensive, requiring large amounts of water (e.g., wet washing), adsorbents, and/or chemicals. Refining, in turn, produces substantial amounts of waste and is accompanied by the loss of biodiesel as neutral oil entrained in waste. A wide array of methods and technologies have been developed for industrial oil purification. Successful refining practices minimize waste and limit neutral oil losses. Recent studies have explored the use of adsorbents, solvent purification processes, membrane filtration, as well as novel applications of electrostatic field treatments to remove polar impurities (including free fatty acids, residues, soaps, and glycerides), and particulates from oils. This chapter will review and compare traditional current and novel strategies for refining FAAE for use as biodiesel.

Keywords: biodiesel, fuel, refining strategies, fatty acid alkyl esters, production of biodiesel

1. Introduction

An increase in global energy demand is driving a shift from traditional fossil fuels to renewable and sustainable energy, such as biofuels (e.g., bioethanol and biodiesel). Potentially, the use of renewable fuels can reduce greenhouse gas production, and pollution related to fossil fuel use [1] while also providing a sustainable source of fuel. Collectively, the biofuel markets is expected to reach $245 billion USD by 2027 [2], with biodiesel alone accounting for $73 billion USD by 2030 [3]. The top global producers of biodiesel and bioethanol and their feedstocks are listed in **Tables 1** and **2**, respectively. Currently, the United States, European Union, and Brazil are major global producers for biofuels (bioethanol and biodiesel) [4]. The COVID-19 pandemic led to restrictions to global travel and transport that resulted in a significant global decline in transport fuel [4] consumption, that in turn placed strain on the supply chain [5]. The subsequent lifting of pandemic-related restrictions in 2021–2022 led to recovery of fossil fuel and biofuel markets. The sudden increase in demand led to higher feedstock prices and bottlenecks in supply chains that have in turn increased production costs [4, 5].

Country	Biodiesel Production (Million L for 2019–2021*)	Feedstock
European Union	14,882	Rapeseed oil/Palm oil/Used cooking oils
United States	8905	Soybean oil/Used cooking oils
Indonesia	8476	Palm oil
Brazil	6325	Soybean oil
Argentina	1765	Soybean oil

**Refers to average estimate between 2019 to 2021.*

Table 1.
Top global producers (2021) of biodiesel and the feedstocks used [4].

Country	Ethanol Production (Million L for 2019–2021*)	Feedstock
United States	58,182	Maize
Brazil	32,748	Sugarcane/Maize
China	10,433	Maize/Cassava
European Union	6112	Sugar beet/Wheat/Maize
Thailand	1794	Molasses/Sugarcane/Cassava

**Refers to average estimate between 2019 to 2021.*

Table 2.
Top global producers (2021) of bioethanol and the feedstocks used [4].

Nonetheless, the biofuel market is expected to achieve a compounded annual growth rate (CAGR) of 7.81% by 2027 [2].

Renewable biomass fuels are classified as primary and secondary biofuels. Primary biofuels are traditionally woody or cellulosic plant material and dry animal wastes that are burned for energy [6], while secondary biofuels can be further categorized into three generations dependent on the feedstock material [7]. First generation biofuels include bioethanol produced from carbohydrate rich input materials such as corn and sugar cane, and biodiesel produced from vegetable oils (such as low erucic acid rapeseed oil and soybean oil), and animal fats such as waste cooking grease [6, 8]. A portion of the inputs for production of first-generation biofuels are derived from edible resources. Second generation biofuels include those that utilize non-edible cellulosic feedstock [7] and fuels produced from non-edible oil-rich plant seeds [6, 9]. Next, third generation bioethanol production relies on algal biomass as a sustainable feedstock [7]. There is also interest in investigating the suitability of microalgae for biodiesel production [10]. Recently, fourth generation approaches for bioethanol production that use genetically engineered organisms are being investigated [11]. Similarly, a fourth generation of biodiesel is based genetically engineered organisms [12].

Second-generation biodiesel can contribute to significant reductions in carbon dioxide emissions when compared with fossil diesel sources as the feedstocks used for their production is often considered carbon neutral (e.g., plant-based) [13]. Second-generation biodiesel has also been extensively studied [14] and often utilize fully or partially refined vegetable oils (e.g., soybean oil and rapeseed oils) [15], that must be transesterified using chemical treatments (e.g., acid and/or base catalytic transesterification, with alkali-catalyzed transesterification being the most common [16]), to form fatty acid alkyl esters (FAAE) [15]. Typically, transesterification leads to the production of several contaminants in the FAAE including glycerol, glycerides, soap, alcohols, as well as accumulated moisture due to improper treatment processes or by adsorption of atmospheric moisture during storage [17]. The presence of these contaminants can have detrimental effects on the quality of the FAAE fuel, performance of vehicles being fueled, and storage properties of the fuel including, but not limited to, clogging of fuel filters [18], corrosion of storage containers [19], lower flashpoint [19], and increased deposit formation in the engine fuel system [19]. Therefore, removal of these contaminants is required to ensure quality and utility of the biodiesel product. However, conventional refining methods can require the use of many inputs, high capital cost equipment, and generate waste. For example, wet washing uses heated and softened water [20], while "dry washing" uses adsorbent materials [21] to remove polar contaminants [22]. These washing methods can result in the production of substantial waste [20] and/or require large amounts of expensive absorbent materials [1]. Furthermore, the introduction of water during wet refining processes can lead to the formation of emulsions which leads to FAAE loss [23], but also often requires specialized infrastructure to reduce moisture content in the biodiesel after refining [22].

In addition to wet and dry washing, alternative refining methods using enzymatic processes such as enzymatic degumming [24], physical refining methods [25], and membrane filtration [26] have been explored. Although these methods have been successful in removing contaminants and reducing chemical consumption and byproduct waste, implementation at an industrial scale can be prohibitive due to the cost of expensive reagents and specialized infrastructure. In addition, there are few solvent-stable membranes, and substitution of the solvent during seed oil extraction may be difficult [26]. For example, alcohols exhibit higher latent heat and lower solvent power, compared to hexane thus, requiring a larger amount of solvent to extract seed oil. More recently, newer refining methods (e.g., electrostatic field refinement) [27, 28] have been investigated as approaches to minimize the ecological footprint compared to conventional methods. The purpose of this review of biodiesel production, is to compare approaches of FAAE refining, and explore strategies for refining FAAE for biodiesel production.

2. Production of biodiesel

Most commercial biodiesel is produced by the transesterification of triglycerides with short-chain alcohols (e.g., methanol, ethanol), in the presence of a catalyst [29, 30], yielding glycerin and FAAE [31, 32]. Catalysts can include alkali catalysis (most common approach), acid catalysis, a sequence of acid catalysis followed by alkali catalyst, or a lipase catalysis [33–37] (**Table 3**). After transesterification, the FAAE phase is separated from glycerol and residual catalyst via settling or centrifugation [1].

Free fatty acids (FFA) in an oil or fat feed material can neutralize alkali catalysts and limit the transesterification reaction required for producing FAAE [15] and ultimately affect the yield and quality of biodiesel [38]. Reactions of the FFA in oils with alkaline catalysts forms soaps (e.g., carboxylic acid salts) and water. The presence of soap increases FAAE viscosity, and forms emulsions that can make separation

Catalyst type	Advantages	Disadvantages
Alkali	More effective, less corrosive	Extra alkali required when processing feedstocks with high content of free fatty acids Increased production of soap
Acid	Appropriate for low quality oil feedstocks	Slow process and results in incomplete reactions
Acid-alkali	Less depletion of catalyst, enhanced FAAE yield, generates less contaminants	Can require additional infrastructure and increase processing time
Lipase	Continuous production	Expensive and can be unstable

Table 3.
Advantages and disadvantages of FAAE production methods.

of glycerol from biodiesel difficult [39]. To prevent this, feedstocks containing FFA should be treated prior to the transesterification process. Frequently used methods to decrease FFAs include acidic esterification, enzymatic esterification, and contacting the FAAE phase with alkaline water. Common acid treatments involve the use of sulfuric acid, in the presence of an alcohol such as methanol or ethanol, to reduce FFA content in the feedstock oil [40–45]. Hydrochloric acid and phosphoric acid can also reduce FFA content in cooking oil destined for biodiesel production [46]. Treatments of high FFA oils with lipase as an enzyme catalysts and glycerol can convert FFAs to glycerides, including monoglycerides, diglycerides, and triglycerides giving an overall reduction of FFA content [47]. The decrease in FFAs helps to improve conditions for alkali transesterification reactions [48].

This transesterification of lipids produces FAAE with beneficial physicochemical properties, including lower viscosity, ignition and flash points, making it an suitable biofuel [49]. However, incomplete transesterification can produce unrefined biodiesel with a variety of contaminants including glycerol, mono–/di-glyceride, soap,

Contaminants	Contaminant limits in standardized guidelines	Potential hazards
Moisture	0.05% (by volume; ASTM D6751)	Fuel filter plugging due to microbial growth; corrosion of storage container
Methanol	1. % (by weight; EN 14110)	Lower flashpoint of biodiesel
FFA	0.5 mg/g KOH	Corrosion, low oxidation stability
Metals	5 mg/kg of Class 1 metals (Na + K; EN 14214)	Deposits in the injector, filter blockage, engine weakening
Soap	66 mg/kg when KOH is used as the catalyst (ASTM D6751)	Conferring biodiesel a degree of electric conductivity
Free glycerol	<0.02% (by weight; ASTM D6751)	Forming gum-like deposit around injector tips and valve heads
Total glyceride	<0.24% (by weight; ASTM D6751)	Increased levels of oxidative deposits

Table 4.
Common contaminants in FAAE and the potential hazards [19, 50–53].

alcohol, moisture, and catalyst [29]. The presence of these contaminants in FAAE could significantly affect engine performance, damage fuel storage equipment, and lead to engine failure (**Table 4**). To ensure that biodiesel fuel is suitable for use as a fuel the crude FAAE must be refined to meet standardized guidelines, such as the American Standard for Testing Materials (ASTM D6751) and the European Standard (EN 14214) [54]. Commercial biodiesel refining includes either wet or "dry" washing, however, these processes have significant drawbacks, such as requiring the use of expensive infrastructure and the production of waste products (e.g., wastewater). More recently, sustainable purification alternatives have been explored for biodiesel to maximize outputs, reduce the generation of waste products, and minimize the need for expensive infrastructure.

3. Conventional FAAE refining strategies

After transesterification, conventional crude FAAE refining strategies include the separation of the FAAE phase from glycerol based on differing densities (e.g., gravitational settling, sedimentation, centrifugation, decantation) [15, 29], followed by wet and/or dry washing (**Figure 1**). Wet washing removes polar contaminants (e.g., glycerol, soap, alcohol, and catalyst residue), from crude FAAE using a combination of heat and softened water utilizing the affinity of polar contaminants with water [20, 55]. The washing step is typically repeated until the biodiesel layer becomes clear and water layer becomes transparent [1]. Since this washing process must be repeated

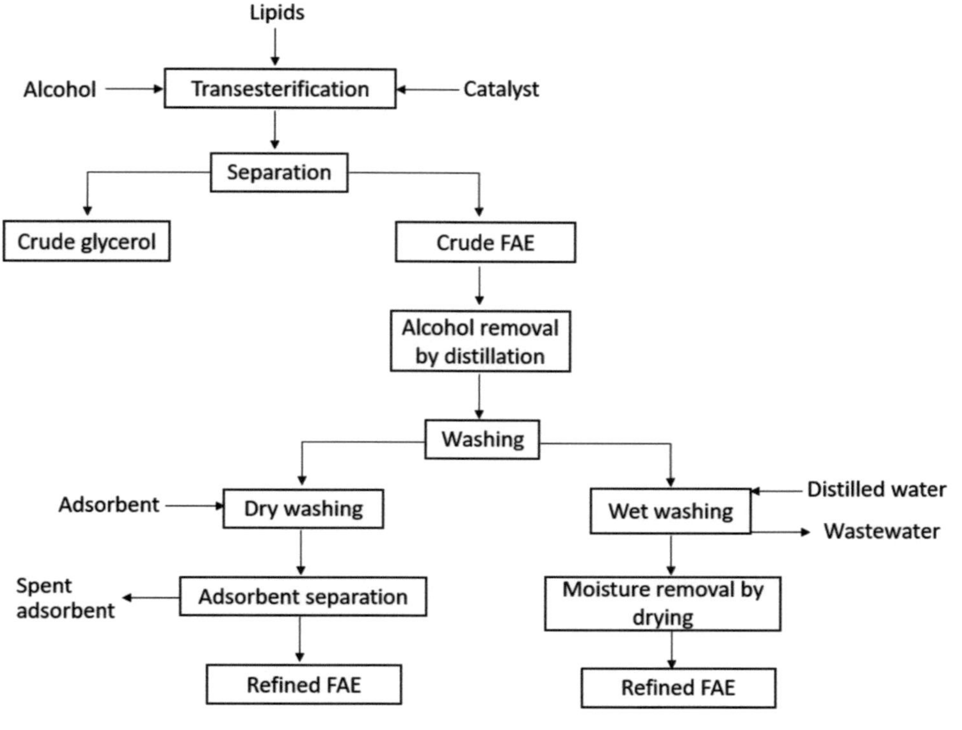

Figure 1.
Typical fatty acid alkyl esters production and purification process.

Refining method	Advantages	Disadvantages
Wet washing	> 99% FAAE purity is achievable	Requires large amounts of water
	Efficient removal of glycerol and methanol	Generates wastewater
	Removes FAAE-soluble impurities	Drying process can be energy intensive
Dry washing	Easier operation and less effort than wet washing	Lower removal rate of contaminants
	Less operation time	Re-generation of adsorbents can be challenging
	No introducing moisture into FAAE	Requires expensive machinery (e.g., pump, column)

Table 5.
Advantages and disadvantages of wet and dry washings in FAAE refining [59].

multiple times, these methods typically require large amounts of water and generate substantial waste in the form of contaminated water [56]. Typically, refining 1 L of crude FAAE can generate between 3 and 10 L of wastewater [29] and an additional drying step must then be included to remove residual moisture and prevent triglyceride hydrolysis [15]. Drying biodiesel can be costly and can result in the formation of emulsions [57].

In comparison, "dry" washing involves the removal of contaminants from crude FAAE through use of adsorbents or ion exchange resins. Suitable adsorbents include silicates (Magnesol™ or TriSyl™), ion exchange resins (Amberlite™ or Purolite™), cellulosic materials, activated clay, activated carbon, activated fiber, and others [29]. "Dry" washing is based on the affinity of adsorbents to common polar contaminants [58]. Although this method does not rely on contacting biodiesel with water, dry washing consumes a considerable amount of adsorbent material. However, this consumption can be mitigated if the adsorbent can be reused [59].

Both wet and dry washing techniques are commonly used for FAAE refining, and the advantages and disadvantages of both methods are summarized in **Table 5**. These conventional washing techniques increase production cost and further contribute to FAAE loss. The use of wet and dry washing to refine the FAAE phase generates substantial waste including wastewater and spent adsorbent. Waste is an important consideration for commercial manufacturers. As a result, there is significant interest in the development of refining technologies that are both efficient and environmentally sustainable as further elaborated below.

4. Novel fatty acid alkyl esters refining methods

4.1 Membrane filtration

Conventional FAAE refining processes have significant drawbacks including high energy usage, oil losses, use of harmful chemicals, and the production of substantial waste or effluent products [26]. Fortunately, membrane technologies offer advantages including customizable features in the process design [60] that is both efficient and environmentally sustainable for the purification and separation of many biofuels, including biodiesel [61]. Some advantages include reducing thermal damage to the end-product, improved solvent recyclability, decreased emissions and energy consumption, minimized oil losses, and reduced bleaching requirements [26]. Membrane processes have been developed for refining biodiesel and such refining can generate high quality FAAE with increased yields when compared with conventional processes [62, 63]. Specifically, membranes with suitable properties (tolerance of mechanical, chemical, and thermal stress, high surface area, enhanced selectivity can remove particles from biodiesel [62]. Micro-filtration (MFM; 0.1–10 mm pore size) and ultra-filtration membranes (UFM; 1–100 nm pore size) have been tested for the removal of particles from crude FAAE [53]. These membranes can be classified as either organic membrane (e.g., cellulose membranes) and inorganic membrane (e.g., ceramic membranes), with the latter having the advantage of increased durability, thermostability, and stability in acid, alkali, and high-pressure environments [64]. Common types of membranes used in FAAE purification are summarized in **Table 6**; the two distinct properties that influence the efficacy of the membranes are pore size and perm-selectivity [61].

Membrane filtration has been employed as a refining step for processing FAAE after transesterification [67]. Membrane based refining can reduce or remove common contaminants [67, 68], as well as residual catalysts [69]. For example, filtration through ceramic [70, 71] membranes reduce the glycerol content of FAAE [68]. Once isolated the glycerol byproduct might then be further purified and converted to value-added chemicals (e.g., solketal and glycerol carbonate) [72]. However, the use

Membranes	Pore size (μm)	Temperature (°C)	Permeate Flux ($L\ m^{-2}\ h^{-1}$)	Pressure	Results	Ref.
Ceramic membrane	0.1	60	300	1.5 bar	Low metal and free glycerol content	[65]
Ceramic membrane	0.05	40	22.17	2 bar	97.5% and 96.6% removal rate of free glycerol and soap content	[62]
Mixed cellulose ester membrane	0.22	Ambient temperature	~130	2 bar	Effect of soap and free glycerin removal	[66]
Poly (ether-sulfone) ultrafiltration membrane	10 kDa	Ambient temperature	~80	4 bar	Effect of soap and free glycerin removal	[66]

Table 6.
Application of membranes in fatty acid alkyl esters refining.

of absorbents is typically required to improve the efficiency of ceramic membranes [71]. In addition, filtration processes have been evaluated for the separation of crude biodiesel from unreacted vegetable oil [73], catalyst [65], soap [65], and methanol [1]. Ceramic or carbon microporous membranes are usually preferred due to resistance to corrosion and degradation in the presence of a base or acid catalyst during transesterification [67, 73, 74]. Filtration through ceramic membranes has been shown to effectively refine biodiesel to produce a product that meets standard specifications [65, 70, 71, 74, 75]. As glycerol and FAAE are immiscible due to their polarity differences, the lipid/oil droplets can be filtered through ceramic membranes owing to their large droplet size compared to the biodiesel [61]. These membranes have also been successful removing other common contaminants such as soap, to reach the quality defined in standard specifications [76]. However, it is often necessary to clean the membranes after each purification process either physically, chemically (hydrogen peroxide, chlorine, etc.), or hydraulically [77], to extend the reusability, efficiency, and repeatability of the membrane [76], otherwise, fouling and clogging can occur [74]. Furthermore, the membrane performance can be affected by physical operating parameters, including temperature requirements. Depending on membrane tolerance of operating conditions scale-up strategy and process design can be limited [76]. Computational modeling has also shown promising results in utilizing reverse osmosis filtration to remove impurities present in crude biodiesel [78]. Finally, the use of coagulation and flocculation processes [79], and membrane filtration methods, including nanofiltration, have been explored to treat biodiesel wastewater [80, 81], accumulated during wet washing processes [80].

Implementation of membrane filtration for biodiesel refining provides an environmentally valuable alternative to conventional biodiesel refining strategies [82]. Membrane filtration has demonstrated success in refining biodiesel from different feedstocks and produced products that meet the ASTM D6751 standard [68]. The combination of electrostatic strategies (electrostatic field, electrodialysis, etc.), with membrane filtration technologies, have also been explored and demonstrated high efficiency and success in refining biodiesel [83], and is further elaborated below.

4.2 Application of electrostatic field and nano-adsorbents in FAAE refining

The utilizing of static electric fields as a method for removing contaminants from FAAE has recently been reported [84, 85]. This method uses the behavior of particles in static electric fields [86] to remove them from FAAE. The process separates particles using electrochemical properties that separate them from bulk solution by electrophoresis (EP) and dielectrophoresis (DEP) [87]. Using EP, particles respond to a uniform direct current voltage that attracts particles, while DEP manipulates particles in a non-uniform electric field which can be tailored depending on the medium and contaminant properties [88]. Based on a range of dielectric properties of targeted molecules/particles/contaminants of interest DEP is commonly used in medical applications, diagnostics, environmental research, polymer research, and particle filtration [89]. These principles can be similarly applied for particle removal enabling partial FAAE refining.

Currently, processes involving electric fields have been applied in oil processing as a means for demulsification and dehydration [90] and removing contaminants [91]. These techniques can be employed to effectively remove polar contaminants (e.g., soap, and glycerides) [92–95] from non-polar fluids with dielectric constants below 2.8 (**Table 7**) [99]. The application of electrostatic fields can remove free glycerol from crude FAAE [100–102].

Source	Dielectric constant (k)	Temperature (°C) and frequency (MHz)	Reference
Soybean oil based	3–3.24	25–70, 1 kHz- 100 kHz	[96]
Soybean, sunflower, corn, and rapeseed based	3–3.4	25–75, 20–20 MHz	[97]
Soybean oil based	2.08–2.26	25 100–2000 Hz	[98]

Table 7.
Dielectric constant of vegetable oil based FAAE.

More recently, electrostatic field treatments have been successfully employed in removing polar contaminants present in crude canola oil [27]. Reductions of 74.9%, 53.2%, and 47.0% in phospholipid, free fatty acid, and peroxide contents, respectively, were achieved using a commercial electrostatic "oil cleaner" with a fixed flow rate and equipped with pleated cellulose collector (designed to collect oil contaminants) [27]. Neutral oil loss was minimal (0.37 wt %), and pigment content (e.g., chlorophyll and carotenoid content) remained unaffected [27]. However, it should be noted that the limitations of this process included a requirement to reduce FAME moisture content prior to treatment.

To remove more polar compounds Zhou et al. [27] coupled the use of nano adsorbents (e.g., Al_2O_3, SiO_2, and TiO_2) with electrostatic separation. This implementation of nano-adsorbents coupled with electrostatic field treatment effected a significant reduction in contaminants (e.g., soap, total glycerides) from crude FAME. These reductions were observed after only 1 minute of adsorption and at low agitation speeds [28]. Although removal of spent nano-adsorbents can be challenging, electrostatic field application was successful in removing these particulates from the crude FAME solution without altering FAME chemistry [28]. Unfortunately, the application of some of these nano-adsorbents were limited by their reusability. Only Al_2O_3 was repeatedly regenerated after multiple cycles [28].

E-field and nano-adsorbent treatments offer novel approaches that can be rapidly implemented to process and refine vegetable oil and FAAE for biodiesel production [27, 28]. The ability to reuse spent nano-adsorbents, and potentially recycle solvents [26], make these approaches an attractive alternative to conventional refining. Furthermore, implementation of E-field technologies can offer a greener approach to biodiesel refining as well as economic and environmental advantages including reducing the need for expensive infrastructure, equipment, or harmful chemicals and reducing waste. Further studies should focus on the standardization and optimization of these processes at an industrial scale.

5. Conclusion

The transition from fossil fuels to a new fuel source is becoming increasingly important, and the environmental impact from the production of a new generation of fuel must be carefully considered as the market grows. The use of biofuels is unique in that it uses what can be classified as carbon neutral sources and implements feedstock which can come from primary or secondary sources. The use of these biofuels, however, requires refining which can be costly, environmentally harmful, and energy intensive. Conventional techniques for the removal of contaminants, such as dry and wet washing, produce large amounts of waste. In addition, enzymatic processes, physical refining strategies (temperature and pressure treatments), and membrane filtration can be limited in their capacity due to expensive reagents and infrastructure requirements, or solvent compatibilities. Fortunately, new refinement methods, such electrostatic field treatment and nanoparticle adsorbents have been developed to improve efficiency, minimize chemical usage, and reduce the generation of waste byproducts. These methods have successfully demonstrated the ability to remove contaminants including soap, methanol, and glycerides from vegetable oils and FAAE. Compared to conventional methods, these new techniques offer an economical and environmentally sustainable approach to current conventional methods. These techniques could be considered at an industrial scale, although further optimization is needed. As the global economy's dependence on fossil fuels continues to decrease, biofuels will play an important role in the development of environmentally friendlier materials to ensure that supply chain needs are met.

Acknowledgements

This work was supported by the Saskatchewan Agricultural Development Fund (20190155, 20190154, 20180281, 20180248, 20180255, 20170133); Natural Sciences and Engineering Research Council of Canada Discovery Grant (RGPIN-2018-06631); and Mitacs (IT19122, IT16156).

Conflict of interest

Dr. Martin J.T. Reaney is the founder of, and has an equity interest in, Prairie Tide Diversified Inc. (PTD, Saskatoon, SK, Canada: previous company name is Prairie Tide Chemicals Inc.).

Author details

Timothy J. Tse, Li Zhou, Farley Chicilo, Venkatesh Meda* and Martin J.T. Reaney*
University of Saskatchewan, Saskatoon, Canada

*Address all correspondence to: venkatesh.meda@usask.ca
and martin.reaney@usask.ca

References

[1] Atadashi IM, Aroua MK, Aziz ARA, Sulaiman NMN. Refining technologies for the purification of crude biodiesel. Applied Energy. 2011;**88**(12):4239-4251. DOI: 10.1016/J.APENERGY.2011.05.029

[2] Market Research Report. 2022. Biofuels market research report: Information, by fuel type (biodiesel and ethanol), by feedstock type (first, second, and third generation), and by region (North America, Europe, Asia-Pacific, the Middle East & amp; Africa, and South America) - Forecast till. 2030

[3] Grand View Research. Biodiesel market size, share & trends analysis report by feedstock (vegetable oils, aniimal fats), by application (fuel, power generation), by region (Europe, APAC), and segment forecasts, 2022-2030; grand view forecasts, 2020-2027: San Francisco, CA, USA. 2022

[4] OECD-FAO Agricultural outlook 2022-2031. 2022

[5] Mustafa R, Purdy SK, Nelson FB, Tse TJ, Wiens DJ, Shen J, et al. World Medical & Health Policy. 2021:1-13. DOI: 10.1002/wmh3.463

[6] Rodionova MV, Poudyal RS, Tiwari I, Voloshin RA, Zharmukhamedov SK, Nam HG, et al. Biofuel production: Challenges and opportunities. International Journal of Hydrogen Energy. 2017;**42**:8450-8461. DOI: 10.1016/j.ijhydene.2016.11.125

[7] Tse TJ, Wiens DJ, Reaney MJT. Production of bioethanol - A review of factors affecting ethanol yield. Fermentation. 2021;7(4):268. DOI: 10.3390/fermentation7040268

[8] Saeid B, Aroua MK, Abdul Raman A, Sulaiman NMN. Density of palm oil-based methyl ester. Journal of Chemical and Engineering Data. 2008;**53**:877-880

[9] Qiu F, Li Y, Yang D, Li X, Sun P. Biodiesel production from mixed soybean oil and rapeseed oil. Applied Energy. 2011;**88**:2050-2055

[10] Helena MA, Catarina GA, Xavier MF. Advances and perspective in using microalgae to produce biodiesel. Applied Energy. 2011;**88**:3402-3410

[11] Ziolkowska JR. Biofuels technologies. In: Biofuels for a More Sustainable Future. Amsterdam, The Netherlands: Elsevier Inc.; 2020. pp. 1-19

[12] Singh D, Sharma D, Soni SL, Inda CS, Sharma S, Sharma PK, et al. A comprehensive review on 1st-generation biodiesel feedstock palm oil: Production, engine performance, and exhaust emissions. Bioenergy Research. 2021;**14**: 1-22. DOI: 10.1007/s12155-020-10171-2

[13] Jeswani HK, Azapagic A. Life cycle sustainability assessment of second generation biodiesel. In: Luque R, Melero JA, editors. Advances in Biodiesel Production. 1st ed. Sawston, UK: Woodhead Publishing; 2012. pp. 13-31. DOI: 10.1533/9780857095862.1.13

[14] Pinzi S, Dorado MP. Feedstocks for advanced biodiesel production. In: Luque R, Melero JA, editors. Advances in Biodiesel Production. Cambridge, UK: Woodhead Publishing Series in Energy; 2012:69-90. DOI: 10.1533/978085709862.1.69

[15] Elgharbawy AS, Sadik WA, Sadek OM, Kasabay MA. A review on biodiesel feedstocks and production

technologies. Journal of the Chilean Chemical Society. 2021;**66**:5098-5109. DOI: 10.4067/S0717-97072021000105098

[16] Díaz L, Mertes L, Brito A, Rodríguez KE. Valorization of energy crop shells as potential green adsorbents for free fatty acid removal from oils for biodiesel production. Biomass Conversion and Biorefinery. 2020;**12**: 1-14. DOI: 10.1007/S13399-020-0189-Y

[17] He BB, Thompson JC, Routt DW, Van Gerpen JH. Moisture absorption in biodiesel and its petro-diesel blends. Applied Engineering in Agriculture. 2007;**23**:71-76

[18] Chesneau HL. Fuel system contaminants: An introduction, from distillate fuel: Contamination, storage and handling. In: Chesneau HL, Dorris MM, editors. ASTM Publication STP 1005. Philadelphia, PA: American Society for Testing and Materials; 1988

[19] Van Gerpen JH, Hammond EG, Yu L, Monyem A. Determining the influence of contaminants on biodiesel properties. SAE Technical Papers. 1997. DOI: 10.4271/971685

[20] Abbaszadeh A, Ghobadian B, Najafi G, Yusaf T. An experimental investigation of the effective parameters on wet washing of biodiesel purification. International Journal of Automotive and Mechanical Engineering. 2014;**9**(1):1525-1537. DOI: 10.15282/IJAME.9.2013.4.0126

[21] Faccini CS, Cunha MED, Moraes MSA, Krause LC, Manique MC, Rodrigues MRA, et al. Dry washing in biodiesel purification: A comparative study of adsorbents. Journal of the Brazilian Chemical Society. 2011;**22**:558-563. DOI: 10.1590.S0103-50532011000300021

[22] Suthar K, Dwivedi A, Joshipura M. A review on separation and purification techniques for biodiesel production with special emphasis on Jatropha oil as a feedstock. Asia Pacific Journal of Chemical Engineering. 2019;**14**:e2361. DOI: 10.1002/APJ.2361

[23] Atadashi IM, Aroua MK, Abdul Aziz AR, Sulaiman NMN. The effects of water on biodiesel production and refining technologies: A review. Renewable and Sustainable Energy Reviews. 2012;**16**:3456-3470. DOI: 10.1016/j.rser.2012.03.004

[24] Clausen K. Enzymatic oil-degumming by a novel microbial phospholipase. European Journal of Lipid Science and Technology. 2001;**103**:333-340. DOI: 10.1002/1438-9312(200106)103:6<333:AID-EJLT333>3.0.CO;2-f

[25] Prieto MM, Bada JC, Léon-Camacho M, Constante EG. Deacidification and recovery of distillates in the physical refining of edible oils. European Journal of Lipid Science and Technology. 2008;**111**:101-110. DOI: 10.1002/ejlt.200700149

[26] Cheryan M. Membrane technology in the vegetable oil industry. Membrane Technology. 2005;**2005**:5-7. DOI: 10.1016/S0958-2118(05)70387-6

[27] Zhou L, Shen J, Tse TJ, Chicilo F, Purdy SK, Meda V, et al. Electrostatic field treatment as a novel and efficient method for refining crude canola oil. Journal of Cleaner Production. 2022;**360**:131905. DOI: 10.1016/j.jclepro.2022.131905

[28] Zhou L, Tse TJ, Chicilo F, Shen J, Meda V, Reaney MJT. Electrostatic field and nano-adsorbent refining of fatty acid methyl esters. Journal of Cleaner Production. 2022;**373**:133679. DOI: 10.1016/j.jclepro.2022.133679

[29] Atadashi IM, Aroua MK, Abdul Aziz AR, Sulaiman NMN. Membrane biodiesel production and refining technology: A critical review. Renewable and Sustainable Energy Reviews. 2011;**15**(9):5051-5062. DOI: 10.1016/J.RSER.2011.07.051

[30] Musa IA. The effects of alcohol to oil molar ratios and the type of alcohol on biodiesel production using transesterification process. Egyptian Journal of Petroleum. 2016;**25**(1):21-31. DOI: 10.1016/J.EJPE.2015.06.007

[31] Haas MJ, McAloon AJ, Yee WC, Foglia TA. A process model to estimate biodiesel production costs. Bioresource Technology. 2006;**97**(4):671-678. DOI: 10.1016/J.BIORTECH.2005.03.039

[32] Hirner FS, Hwang J, Bae C, Patel C, Gupta T, Agarwal AK. Performance and emission evaluation of a small-bore biodiesel compression-ignition engine. Energy. 2019;**183**:971-982. DOI: 10.1016/J.ENERGY.2019.07.015

[33] Jegannathan KR, Eng-Seng C, Ravindra P. Economic assessment of biodiesel production: Comparison of alkali and biocatalyst processes. Renewable and Sustainable Energy Reviews. 2011;**15**(1):745-751. DOI: 10.1016/J.RSER.2010.07.055

[34] Mansir N, Taufiq-Yap YH, Rashid U, Lokman IM. Investigation of heterogeneous solid acid catalyst performance on low grade feedstocks for biodiesel production: A review. Energy Conversion and Management. 2017;**141**:171-182. DOI: 10.1016/J.ENCONMAN.2016.07.037

[35] Meng X, Chen G, Wang Y. Biodiesel production from waste cooking oil via alkali catalyst and its engine test. Fuel Processing Technology. 2008;**89**(9):851-857. DOI: 10.1016/J.FUPROC.2008.02.006

[36] Tan T, Lu J, Nie K, Deng L, Wang F. Biodiesel production with immobilized lipase: A review. Biotechnology Advances. 2010;**28**(5):628-634. DOI: 10.1016/J.BIOTECHADV.2010.05.012

[37] Zhang J, Chen S, Yang R, Yan Y. Biodiesel production from vegetable oil using heterogenous acid and alkali catalyst. Fuel. 2010;**89**(10):2939-2944. DOI: 10.1016/J.FUEL.2010.05.009

[38] Bouaid A, Vázquez R, Martinez M, Aracil J. Effect of free fatty acids contents on biodiesel quality. Pilot plant studies. Fuel. 2016;**174**:54-62. DOI: 10.1016/j.fuel.2016.01.018

[39] Ramadhas AS, Jayaraj S, Muraleedharan C. Biodiesel production from high FFA rubber seed oil. Fuel. 2005;**84**:335-340. DOI: 10.1016/j.fuel.2004.09.016

[40] Hayyan A, Alam MZ, Mirghani MES, Kabbashi NA, Hakimi NINM, Siran YM, et al. Reduction of high content of free fatty acid in sludge palm oil via acid catalyst for biodiesel production. Fuel Processing Technology. 2011;**92**:920-924. DOI: 10.1016/j.fuproc.2010.12.011

[41] Chai M, Tu Q, Lu M, Yang J. Esterification pretreatment of free fatty acid in biodiesel production from laboratory to industry. Fuel Processing Technology. 2014;**125**:106-113. DOI: 10.1016/j.fuproc.2014.03.025

[42] Kara K, Ouanji F, Lotfi EM, Mahi ME, Kacimi M, Ziyad M. Biodiesel production from waste fish oil with high free fatty acid content from Moroccan fish-processing industries. Egyptian Journal of Petroleum. 2018;**27**:249-255. DOI: 10.1016/j.ejpe.2017.07.010

[43] Abdullah, Sianipar RNR, Ariyani D, Nata IF. Conversion of palm oil sludge

to biodiesel using alum and KOH as catalysts. Sustainable Environment Research. 2017;**27**:291-295. DOI: 10.1016/j.serj.2017.07.002

[44] Marchetti JM, Errazu AF. Esterification of free fatty acids using sulfuric acid as catalyst in the presence of triglycerides. Biomass and Bioenergy. 2008;**32**:892-895. DOI: 10.1016/j.biombioe.2008.01.001

[45] Thaiyasuit P, Pianthong K, Worapun I. Acid esterification - Alkaline transesterification process for methyl ester production from crude rubber seed oil. Journal of Oleo Science. 2012;**61**: 81-88. DOI: 10.5650/jos.61.81

[46] Sahar, Sadaf S, Iqbal J, Ullah I, Bhatti HN, Nouren S, et al. Biodiesel production from waste cooking oil: An efficient technique to convert waste into biodiesel. Sustainable Cities and Society. 2018;**41**:220-226. DOI: 10.1016/j.scs.2018.05.037

[47] Fregolente PBL, Fregolente LV, Pinto GMF, Batistella BC, Wolf-Maciel MR, Filho RM. Monoglycerides and diglycerides synthesis in a solvent-free system by lipase-catalyzed glycerolysis. Applied Biochemistry and Biotechnology. 2008;**146**:165-172. DOI: 10.1007/s12010-008-8133-3

[48] Kombe GG, Temu AK, Rajabu HM, Mrema GD, Kansedo J, Lee KT. Pre-treatment of high free fatty acids oils by chemical re-esterification for biodiesel production. Advances in Chemical Engineering and Science. 2013;**3**:242-247. DOI: 10.4236/aces.2013.34031

[49] Candeia RA, Silva MCD, Carvalho Filho JR, Brasilino MGA, Bicudo TC, Santos IMG, et al. Influence of soybean biodiesel content on basic properties of biodiesel–diesel blends. Fuel. 2009;**88**(4):738-743. DOI: 10.1016/J.FUEL.2008.10.015

[50] Berrios M, Skelton RL. Comparison of purification methods for biodiesel. Chemical Engineering Journal. 2008;**144**(3):459-465. DOI: 10.1016/J.CEJ.2008.07.019

[51] Van Gerpen J. Biodiesel processing and production. Fuel Processing Technology. 2005;**86**(10):1097-1107. DOI: 10.1016/J.FUPROC.2004.11.005

[52] McCormick RL, Ratcliff M, Moens L, Lawrence R. Several factors affecting the stability of biodiesel in standard accelerated tests. Fuel Processing Technology. 2007;**88**(7):651-657. DOI: 10.1016/J.FUPROC.2007.01.006

[53] Saleh J, Tremblay AY, Dubé MA. Glycerol removal from biodiesel using membrane separation technology. Fuel. 2010;**89**(9):2260-2266. DOI: 10.1016/J.FUEL.2010.04.025

[54] Jamaluddin NAM, Riayatsyah TMI, Silitonga AS, Mofijur M, Shamsuddin AH, Ong HC, et al. Techno-economic analysis and physicochemical properties of ceiba pentandra as second-generation biodiesel based on ASTM D6751 and EN 14214. PRO. 2019, 2019;7(9):636. DOI: 10.3390/PR7090636

[55] Low SC, Gan GK, Cheong KT. Separation of methyl ester from water in a wet neutralization process. Journal of Sustainable Energy & Environment. 2011;**2**:15-19

[56] Veljković VB, Stamenković OS, Tasić MB. The wastewater treatment in the biodiesel production with alkali-catalyzed transesterification. Renewable and Sustainable Energy Reviews. 2014;**32**:40-60. DOI: 10.1016/J.RSER.2014.01.007

[57] El-Mashad HM, Zhang R, Avena-Bustillos RJ. A two-step process for biodiesel production from

salmon oil. Biosystems Engineering. 2008;**99**(2):220-227. DOI: 10.1016/J.BIOSYSTEMSENG.2007.09.029

[58] Gomes MG, Pasquini D. Utilization of eggshell waste as an adsorbent for the dry purification of biodiesel. Environmental Progress & Sustainable Energy. 2018;**37**(6):2093-2099. DOI: 10.1002/EP.12870

[59] Stojković IJ, Stamenković OS, Povrenović DS, Veljković VB. Purification technologies for crude biodiesel obtained by alkali-catalyzed transesterification. Renewable and Sustainable Energy Reviews. 2014;**32**:1-15. DOI: 10.1016/J.RSER.2014.01.005

[60] Kumar R, Ghosh AK, Pal P. Sustainable production of biofuels through membrane-integrated systems. Separation and Purification Reviews. 2019;**49**:1-22. DOI: 10.1080/15422119/2018/1562942

[61] Hajilary N, Rezakazemi M, Shirazian S. Biofuel types and membrane separation. Environmental Chemistry Letters. 2019;**17**:1-18. DOI: 10.1007/s10311-018-0777-9

[62] Atadashi IM, Aroua MK, Abdul Aziz AR, Sulaiman NMN. Crude biodiesel refining using membrane ultra-filtration process: An environmentally benign process. Egyptian Journal of Petroleum. 2015;**24**(4):383-396. DOI: 10.1016/J.EJPE.2015.10.001

[63] Atadashi IM, Aroua MK, Aziz AA. Biodiesel separation and purification: A review. Renewable Energy. 2011, 2011;**2**:437-443. DOI: 10.1016/j.renene.2010.07.019

[64] Saracco G, Neomagus HWJP, Versteeg GF, Van Swaaij WPM. High-temperature membrane reactors: Potential and problems. Chemical Engineering Science. 1999;**54**(13-14):1997-2017. DOI: 10.1016/S0009-2509(99)00009-3

[65] Wang Y, Xingguo W, Yuanfa L, Shiyi O, Yanlai T, Shuze T. Refining of biodiesel by ceramic membrane separation. Fuel Processing Technology. 2009;**90**:422-427. DOI: 10.1016/j.fuproc.2008.11.004

[66] Alves MJ, Nascimento SM, Pereira IG, Martins MI, Cardoso VL, Reis M. Biodiesel purification using micro and ultrafiltration membranes. Renewable Energy. 2013;**58**:15-20. DOI: 10.1016/J.RENENE.2013.02.035

[67] Baroutian S, Aroua MK, Raman AAA, Sulaiman NMN. A packed bed membrane reactor for production of biodiesel using activated carbon supported catalyst. Bioresource Technology. 2011;**102**:1096-1102. DOI: 10.1016/j.biortech.2010.08.076

[68] Cao P, Dubé MA, Tremblay AY. High-purity fatty acid methyl ester production from canola, soybean, palm, and yellow grease lipids by means of a membrane reactor. Biomass and Bioenergy. 2008;**32**:1028-1036. DOI: 10.1016/j.biombioe.2008.01.020

[69] Atadashi IM, Aroua ML, Abdul Aziz AR, Sulaiman NMN. Removal of residual palm oil-based biodiesel catalyst using membrane ultra-filtration technique: An optimization study. Alexandria Engineering Journal. 2014;**53**:705-715. DOI: 10.1016/j.aej.2014.07.002

[70] Gomes MCS, Pereira NC, de Barros STD. Separation of biodiesel and glycerol using ceramic membranes. Journal of Membrane Science. 2010;**352**:271-276. DOI: j.memsci.2010.02.030

[71] Goswami KP, Pugazchenthi G. Effect of binder concentration on properties of low-cost fly ash-based tubular ceramic membrane and its application in separation of glycerol from biodiesel. Journal of Cleaner Production. 2021;**319**:128679. DOI: 10.1016/j.clepro.2021.128679

[72] Chol CG, Dhabhai R, Dalai AK, Reaney M. Purification of crude glycerol derived from biodiesel production process: Experimental studies and techno-economic analyses. Fuel Processing Technology. 2018;**178**:78-87. DOI: 10.1016/j.fuproc.2018.05.023

[73] Dubé MA, Tremblay AY, Liu J. Biodiesel production using a membrane reactor. Bioresource Technology. 2007;**98**:639-647. DOI: 10.1016/j.biortech.2006.02.019

[74] Saleh J, Dubé MA, Tremblay AY. Separation of glycerol from FAME using ceramic membranes. Fuel Processing Technology. 2011;**92**:1305-1310. DOI: 10.1016/j.fuproc.2011.02.005

[75] Atadashim IM, Aroua MK, Abdul Aziz AR, Sulaiman NMN. High quality biodiesel obtained through membrane technology. Journal of Membrane Science. 2012;**412-422**:154-164. DOI: 10.1016/j.memsci.2012.07.006

[76] Ostojčić M, Brkić S, Tišma M, Zelić B, Budžaki S. Membrane filtration as an environmentally friendly method for crude biodiesel purification. Kemija u Industriji. 2019;**69**:175-181. DOI: 10.15255/KUI.2019.049

[77] Atadashi IM. Purification of crude biodiesel using dry washing and membrane technologies. Alexandria Engineering Journal. 2015;**54**:1265-1272. DOI: 10.1016/j.aej.2015.08.005

[78] Ramanan CJ, Garg S, Bora BJ, Buradi A, Roy S, Sharma P, et al. Influence of pressure in RO filtration of crude biodiesel. Conference Series: Earth and Environmental Science. 2022;**1074**:012014. DOI: 10.1088/1755-1315/1074/1/012014

[79] Gonçalves BR, Machado AEH, Trovó AG. Treatment of a biodiesel effluent by coupling coagulation-flocculation, membrane filtration and Fenton reactions. Journal of Cleaner Production. 2016;**142**:1918-1921. DOI: 10.1016/j.jclepro.2016.11.092

[80] Torres JJ, Cuello M, Ochoa NA, Pagliero C. Biodiesel wastewater treatment using nanofiltration membranes. Process Safety and Environmental Protection. 2021;**148**:825-833. DOI: 10.1016/j.psep.2021.02.013

[81] Delcolle R, Gimenes ML, Fortulan CA, Moreira WM, Martins ND, Pereira NC. A comparison between coagulation and ultrafiltration processes for biodiesel wastewater treatment. Chemical Engineering Transactions. 2017;**57**:271-276. DOI: 10.3303/CET1757046

[82] Cheryan M. Membrane Technology in the Vegetable Oil Industry. IL, USA: Agricultural Bioprocess Laboratory, University of Illinois at Urbana - Champaign; 2005

[83] Sdrula N. A study using classical or membrane separation in the biodiesel process. Desalination. 2010;**250**:1070-1072. DOI: j.desal.2009.09.110

[84] Guo Q, Lei C, Chen W, Zhang J, Huang B. Electric-field-driven nanoparticles produce dual-functional bipolar electrodes and nanoelectrolytic cells for water remediation. Cell Reports Physical Science. 2021;**2**(1):100299. DOI: 10.1016/J.XCRP.2020.100299

[85] Xu X, Deng F, Shao P, Dionysiou DD, Luo X, Li X, et al. Internal electric field

driving separation and migration of charge carriers via Z-scheme path in AgIn5S8/ZnO heterojunction for efficient decontamination of pharmaceutical pollutants. Chemical Engineering Journal. 2022;**428**:132096. DOI: 10.1016/J.CEJ.2021.132096

[86] Arutunian SG, Aginian MA, Margaryan AV, Lazareva EG, Chung M. Electric field lines of an arbitrarily moving charged particle. 2021. DOI: 10.48550/arxiv.2109.10792

[87] Pysher MD, Hayes MA. Electrophoretic and dielectrophoretic field gradient technique for separating bioparticles. Analytical Chemistry. 2007;**79**(12):4552-4557. DOI: 10.1021/AC070534J

[88] Rahman NA, Ibrahim F, Yafouz B. Dielectrophoresis for biomedical sciences applications: A review. Sensors. 2016;**17**:449. DOI: 10.3390/s17030449

[89] Riteshkumar A, Komal H, Sankaranarayanan A, Krishnamurthy R. Applications of dielectrophoresis in the field of medical sciences. International Journal of Scientific Research in Science and Technology. 2019;**4**:328-341. DOI: 10.32628/IJSRST18401161

[90] Gong H, Yu B, Dai F, Peng Y, Shao J. Simulation on performance of a demulsification and dewatering device with coupling double fields: Swirl centrifugal field and high-voltage electric field. Separation and Purification Technology. 2018;**207**:124-132. DOI: 10.1016/J.SEPPUR.2018.06.049

[91] Sasaki A, Uchiyama S. A new technology for oil management: electrostatic oil cleaner. SAE Transcations. 2002;**111**:76-82

[92] Brew CR, O'Dowd DJ, Rae ID. Seed dispersal by ants: Behaviour-releasing compounds in elaiosomes. Oecologia. 1989;**80**(4):490-497. DOI: 10.1007/BF00380071

[93] Demafelis RB. Potential of Biodiesel Production from Jatropha Curcas in the Philippines. Laguna, Philippines: SEARCA Professorial Chair Lecture. College; 2007. Available from: https://agris.fao.org/agris-search/search.do?recordID=PH2008000635. [Accessed 11 02 2023]

[94] Dossat V, Combes D, Marty A. Lipase-catalysed transesterification of high oleic sunflower oil. Enzyme and Microbial Technology. 2002;**30**(1):90-94. DOI: 10.1016/S0141-0229(01)00453-7

[95] Hofmann AF. A physicochemical approach to the intraluminal phase of fat absorption. Gastroenterology. 1966;**50**(1):56-64. DOI: 10.1016/S0016-5085(66)80101-4

[96] Corach J, Galván EF, Sorichetti PA, Romano SD. Estimation of the composition of soybean biodiesel/soybean oil blends from permittivity measurements. Fuel. 2019;**235**:1309-1315. DOI: 10.1016/J.FUEL.2018.08.114

[97] Prieto LEG, Sorichetti P, Romano SD. Electric properties of biodiesel in the range from 20 to 20 MHz. Comparison with diesel fossil fuel. International Journal of Hydrogen Energy. 2008;**33L**:3531-3537. DOI: 10.1016/j.ijhydene.2007.10.036

[98] Sander A, Antonije Koščak M, Kosir D, Milosavljević N, Parlov Vuković J, Magić L. The influence of animal fat type and purification conditions on biodiesel quality. Renewable Energy. 2018;**118**:752-760. DOI: 10.1016/J.RENENE.2017.11.068

[99] Aono M, Nitta S, Iwasaki T, Yokoi H, Itoh T, Nomomura S. Amorphous carbon

nitride films as a candidate for low dielectric constant materials. MRS Online Proceedings Library. 1999;**565**:291-296. DOI: 10.1557/PROC-565-291

[100] Abbaszadeh A, Ghobadian B, Najafi G. High voltage/low amperage current for separation of crude glycerin from biodiesel. Advances in Environmental Biology. 2013;**15**:4028-4032

[101] Hinthao T, Wongwuttanasatian T, Kampeerawat W, Suksri A. Glycerin separation from biodiesel transesterification process by pulsed electric field with specific pulse forming network. IOP Conference Series: Materials Science and Engineering. 2020;**859**(1):012012. DOI: 10.1088/1757-899X/859/1/012012

[102] Mayvan AA, Ghobadian B. Electrostatic coagulation for separation of crude glycerin from biodiesel development of a solar powered membrane capacitive deionization system view project a hybrid FSAW-LCSA model development for gasification of sugar cane bagasse view project. 2014

Section 5

Biodiesel Life Cycle

Chapter 8

Evaluation Trials and Carbon Sequestration Potential of *Jatropha curcas* and *Pongamia pinnata*: Technologies and Way Forward

Rajeshwar Rao Gandhe, Ajin Sekhar and Rajkumar Muthu

Abstract

The Chapter focuses on two crops namely *Pongamia pinnata* and *Jatropha curcas*, their germplasm collection, evaluation trials including progeny trials, identification of superior germplasm for maximum yield of oil per unit area, mass multiplication, on-farm trials, carbon sequestration capacity, and successful agro-forestry models. Since India's edible oil consumption needs are heavily dependent on imports, the only feasible way to augment biofuel production is through utilisation of non -edible tree borne oils. Indian demography (population size and population density) coupled with food-fuel competition warrants the use of only wastelands for cultivation with crops tolerant/resilient to severe environmental stress. *P. pinnata* and *J. curcas* are proven, ideal candidates that fit in the narrative and hence the chapter encompasses a holistic, multi-dimensional approach on biodiesel production technologies using *P. pinnata* and *J. curcas* and along with their future prospects.

Keywords: biodiesel, production technologies, tree-borne oilseeds (TBOs), agroforestry, carbon sequestration potential

1. Introduction

Global warming and climate change has shifted the energy requirement paradigm from fossil fuels to biofuels. Among the differentbiofuels production technologies, Tree Borne Oilseeds (TBOs) provide a unique production substrate as it possesses 15–65% oil content from seeds of multi-purpose tree species, which are non-edible and can be utilised as a source of bio-diesel production. Moreover, TBOs has enormous potential in ensuring livelihood security because it involves participation of communities in growing trees as a component and processing of seeds for oil extraction. These oil tree species can be grown across the country under different agro-ecological regions ranging from forests, non-forest areas, degraded lands, barren lands, deserts and hilly landscapes.

Although biofuels derived from TBOs have proven to be ideal eco-friendly candidates for energy security, its mass production remains arduous due to several technological and economic issues ranging from commercial production and harvesting of TBOs at optimum maturity index. The primeimpediments in establishment of an integrated system of large-scale TBOs production are the non-availability of a robust marketing chain with horizontal and vertical linkages and lack of government aided monetary incentives.

India has an appreciablerepository of non-edible tree borne oilseeds as well as the required policies that render them amenable to increased biofuels production. There is also a large collection of germplasm accessions for important TBOs such as *Jatropha curcas*, *Azadirachta indica*, *Pongamia pinnata* and *Madhuca longifolia* that are being maintained and evaluated at different research facilities in the country. In the Research and Development (R&D) front, there have been substantial advancements in aspects such as genetic variability at morphological, biochemical and molecular level, diversity, growth pattern, reproductive biology, yield, propagation, crop improvement, carbon sequestration potential and agroforestry opportunities. What remains to be addressed are the problems encountered in TBOs such as seed collection from scattered locations, processing, high dormancy and problems in picking and harvesting in avenue and forest plantations, non-availability of quality planting material or seed, limited period of availability, unreliable and improper marketing channels, lack of post-harvest technologies and their processing, non-remunerative prices, wide gap between potential and actual production, absence of state incentives promoting bio-diesel as fuel, and economics and cost-benefit ratio.

Uncertainties brought about, by climate change as a result of continually rising greenhouse gas emission, coupled with fluctuating price of crude oil, tense relations among global nations, and rising energy needs, there has been a renewed focus on biofuels in recent years. Biofuels are clean, renewable sources of energy [1]. *J. curcas* has always been a top contender among all potential crop candidates for biofuels production since its seeds are abundant in extractable oil and can provide profitable yields in dry to semi-arid environments and on marginal soils [2, 3]. The oil is more efficient than diesel in pure, blended, and biodiesel forms and causes minimum smoke emission or engine damage [4]. *J. curcas* is widely planted in several countries, and commercial cultivation is being considered by many countries and institutions, including the World Bank [3]. The Indian government has also placed a high priority on *J. curcas* research, with plans to plant a large area of the crop in order to increase biofuel production in the coming years.

After being recognised as a greenhouse gas emissions (GHG) mitigation strategy under the Kyoto Protocol, global interest in carbon sequestered by agroforestry systems gained significant momentum. Comprehensive estimates of biomass and carbon stocks in plantations, including trees outside forests (TOFs), are required to prepare national roadmaps as a part of the commitment made under United Nations Framework Convention on Climate Change (UNFCCC). Furthermore, there is a growing interest in the market opportunities for forest carbon credits [5]. As plans are being implemented to expand *J. curcas* cultivation in large areas throughout the world, there is a need to assess the system's carbon sequestration capacity, which has received less attention than its oil production. This highlights the importance of developing methods to reliably assess the biomass production by the *J. curcas* system, as quantification of carbon in trees is heavily reliant on biomass estimation.

Biomass energy is a local energy source that can sufficiently meet the basic necessities of rural households. Though the contribution of biomass sources to the overall energy scenario is gradually decreasing, it still accounts for more than 40% of the country's energy supply. In rural areas, fuel-wood accounts for 65% of biomass energy, agricultural waste accounts for 20%, and cow dung accounts for 15%. With the increased use of commercial energy sources, there has recently been a significant shift towards commercial sources. As a result, future energy projections in India do not show a proportionate increase in fuel-wood consumption with rising population. It is difficult to predict the fuel-mix shift at this juncture, but it is evident that it is happening in the right direction. Furthermore, in terms of global energy policy, the final form of energy is more important than the primary form. As a result, there has been a strong emphasis on how fuel-wood and other sources of energy can be converted into desirable forms, that are economically feasible, environmentally sound and energy efficient. This transformation is gradual but noticeable.

The Panchayat, or local self-government, is a body of elected citizens that work for the developmental goals and aspirations of the country at the grass root level (rural villages). According to the Eleventh Schedule of the Indian Constitution, certain developmental functions have been delegated to the Panchayats. The Panchayats are in charge of social forestry, farm forestry, land improvement, the implementation of land reforms, land consolidation and soil conservation, fuel and feed production, and non-conventional energy sources. To see how biomass production might be better regulated and regularised through local governance systems, one must analyse the energy policy and rural energy planning initiatives as well as the current programmes of the Indian government.

2. Current and future biofuels' demands

2.1 Global scenario

Demand of biofuels in the global market show distinct diversions between pre-pandemic and post pandemic period. COVID 19 pandemic plummeted the use of biofuels by 8.7% in 2020 relative to 2019. International Energy Agency projects global demand of biofuels to grow by 28% during 2021–2026. Conductive national policies, international commitments to adhere to binding agreements on climate and environment followed by ethanol -fuel blending mandates are expected to boost global biofuel market for 2021–2030. The recent ban on palm oil exports imposed by Indonesia, world's largest producer of palm oil, due to surge in domestic price and supply chain shock induced by Russia – Ukraine crisis have far-reaching ramifications in the demand, supply and use of biofuels by world nations. Nevertheless, demand of biofuels is expected to rise in long run due to the pertinent changes caused in the environment due to climate crisis, unfettered urbanisation at the cost of natural resources and socio-economic developmental goals of the nation states. International initiatives such as Roundtable on Sustainable Biomaterials (RSB), Sustainable Biofuels Consensus and Bonsucro also provide promising platforms to help take decisive actions to ensure both trade and use of biofuels.

2.2 State of biofuel requirement in India

India is ranked 3rd in the world for oil consumption, 20th for oil production globally, and imports nearly 87% of its oil consumption from different trade partners. India's crude oil production has been declining consistently since 2011–2012. India's crude oil imports dictated by expensive import bills coupled with national action plan to adhere to usage limits imposed by international climate agreements have righteously streamlined the fuel sector to develop sustainable alternatives to crude oil. Development of Indigenous Cellulolytic Enzyme for the production of biofuels, and development and transferring of 2G Ethanol technology to Oil Marketing Companies (OMCs) by Department of Biotechnology in the Ministry of Science and Technology, Repurpose Used Cooking Oil (RUCO) launched by Food Safety and Standards Authority of India (FSSAI), Ethanol blending policy of the government to achieve 20% ethanol-blending and 5% biodiesel-blending by 2030 and reducing GST on ethanol for blending from 18–5% are steps mooted at different levels in the right direction to facilitate production, supply chain, sustainable trade and use of biofuels in the country. India's requirement of biofuels is henceforth expected to rise significantly for the decades to come.

2.3 National strategies and policies

2.3.1 National mission on oil seeds

India, being the largest consumer of edible oil in the world, imports more than half of its annual edible oil requirements, primarily from Indonesia, Malaysia, Brazil, Argentina, Russia and Ukraine. With the aim of increasing the acreage and production of oil seeds and oil palm in the country, Government of India launched National Edible Oil Mission-Oil Palm (NMEO-OP) in 2014. The centrally sponsored scheme targets to augment palm oil production from 3.1 lakh tonnes (2021) to 11.20 lakh tonnes by 2025–2026 and to 28 lakh tonnes by 2029–2030. The scheme also targets to increase the acreage of oil palm cultivation by an additional 6.5 lakh hectares by 2025–2026. Oil palm is a tropical crop thriving best in alluvial and moist loamy soils rich in organic matter that requires evenly distributed rainfall of around 3000–4000 mm per annum. Hence NMEO-OP has laid special emphasis to increase the acreage and productivity of oil palm in India's north-eastern states and the Andaman and Nicobar Islands due to favourable weather conditions. The mission is expected to copiously reduce our dependence on imports of oil palm and other oil seeds.

2.3.2 National policy on biofuels

National Policy on Biofuels, rolled out in 2018 streamlines India's ambitious target of becoming self-reliant nation at multifarious levels. The policy expands the ambit of substrates that can be used for producing bio-ethanol by permitting the use of sugar containing substances such as sweet sorghum, sugar beet and sugarcane juice, starch containing cassava and corn, broken or damaged food grains, and potatoes, unfit for human conception. It also classifies biofuels into three different categories vis-à-vis "basic biofuels" or "first generation biofuels" (1G) encompassing bioethanol and biodiesel and "advanced biofuels" or "second generation Biofuels" (2G) that includes ethanol, Municipal Solid Waste (MSW) to drop-in fuels, and "third generation (3G) biofuels" consisting of bio-CNG. The policy also makes it amenable, the use of surplus food grains for bio-ethanol production, thereby encouraging various supply chain mechanisms to boost up the production of biofuels. Recent amendments in the National Policy on Biofuels 2018, make it amenable to increase the scope of feed stocks required for production of biofuels and advances the Ethanol Blending Target (EBT) of 20% of petrol containing ethanol by 2025–2026 instead of 2030. Promotion of use of biofuels in the transportation sector, not only reduces nation's crude import bill but also aligns the country in achieving Sustainable Development Goals (SDGs).

2.3.3 National agroforestry policy

Agroforestry is the scientific practice of integrating trees and shrubs in farm lands for increasing productivity, sustainability and diversity of the ecosystem as a whole. The beginning of the new millennium posed innumerable challenges to the development of Agroforestry by way of lack of institutional mechanisms, insufficient market infrastructure and incoherent legal provisions, which catalysed India in becoming the first country in the world to adopt a National Agroforestry Policy in 2014. The policy envisages the establishment of National Agroforestry Board to synergize and coordinate the efforts of different stakeholders at national level and to strengthen the livelihood opportunities of rural households through Agroforestry. The policy also achieves significant momentum as Agroforestry is known for its role in carbon sequestration, and achieve global climate goals.

3. *J. curcas* and *P. pinnata*

J. curcas is a multipurpose shrub belonging to the family of Euphorbiaceae. It is a tropical plant that can be developed as a commercial crop in farms or as a hedge along the edges of fields in regions with low to high rainfall. It is a plant with a wide range of benefits, including its acknowledged potential for the production of industrial biofuels. The plant can be planted as a live fence to enclose or keep out agricultural animals, as well as to manage erosion and reclaim degraded lands [1]. *J. curcas* is a robust and highly adaptable crop that may restore wastelands, including those that are being used for food crops [6]. Jatropha also thrives on marginal soils with low nutrient contents and is well adapted to well-drainedand well-aerated soils [2]. Jatropha is promoted as a carbon sequestering crop in the current climate change scenario. The best choice is to have continued access to oil with the extra benefit of reducing greenhouse gas emissions. Although it is native to tropical America, it is extensively dispersed in semi-cultivated or wild areas throughout Africa, Central and South America, India, and South East Asia [7] and has been naturalised in the introduced environment. The oil from the Jatropha plant burns with a clear, smoke-free flame, hence it finds application in torches and lamps. Since it shares many of the desirable traits as diesel oil, inedible Jatropha vegetable oil is being investigated as a potential substitute. Jatropha has important implications for providing rural areas with energy services and can thus help in the search for workable alternatives to fossil fuels to reduce the amount of greenhouse gases in the atmosphere. These qualities, together with its adaptability, make it crucial for the progress of developing nations [8]. *J. curcas* exhibits a great deal of variety in its growth, production, and quality traits [9].

Arboreal legume *P. pinnata* (L.) Pierre belongs to the Millettieae tribe of the Papilionoideae subfamily. This medium-sized tree, which is native to the Indian subcontinent and south-east Asia, has been successfully introduced to humid tropical areas of the world, as well as in some areas of Australia, New Zealand, China, and the USA. Historically, traditional remedies, animal feed, green manure, wood, fish poison, and fuel have all been produced from this plant in India and its neighbouring regions. More significantly, *P. pinnata* is now acknowledged as a potential oil source for the developing biofuels sector. Currently, the tree crop is widely grown throughout the humid low land tropics, as well as in Florida, Hawaii, Malaysia, Oceania, the Philippines, and the Seychelles [10]. A 10-year-old *P. pinnata* tree has a carbon sequestration capability of 74 kg, according to an estimate from ICRISAT-Hyderabad [11]. Periodic destructive sampling is the most accurate method for measuring and tracking the estimate of an above-ground biomass for a plant stand [12–14]. Unfortunately, it warrants time and labour to cut and weigh enough trees to accurately represent an ecosystem's size and species distribution. Destructive tree harvesting is cumbersome and labour-intensive [15–17]. Hence it necessitates long term R&D projects with regular monitoring and financial assistance.

4. Field experiments

The Hayatnagar Research Farm of the Central Research Institute for Dryland Agriculture (17.27°N latitude, 78.35°E longitude, and approximately 515 m above sea level), Hyderabad in the Southern part of India, served as the site for all experiments relating to morpho- and genetic variability, growth and reproductive biology, yield, crop improvement, and carbon sequestration potential. The semi-arid climate in the experimental sitewas marked by warm summers and moderate winters. The typical maximum air temperature varies between 13.5 and 16.8 degrees Celsius in the winter and 35.6 to 38.6 degrees Celsius in the summer (March, April, and May). The site experiences roughly 746.2 mm of annual long-term precipitation, primarily from June to October. The soil has a medium texture, is red, and has a shallow depth (Typic Haplustalf as per USDA soil classification).

4.1 Morpho- and genetic-variability in *J. curcas*

32 high-yielding candidate plus trees (CPTs) of *J. curcas* from various locations were collected for evaluating genetic association, variability in seed and growth characters, and latitudinal and longitudinal spread between 12°41 and 22°00 E longitude and 77° 00 and 84°40Nlatitude covering 11 locations in an area spread of 150,000 km^2. In the progeny trial, significant trait differences were seen in all of the seed characters, including seed morphology and oil content, as well as in the growth characteristics, including plant height, the ratio of female to male flowers, and seed output.

The broad sense heritability was found to be high overall and exceeded 80% for all the examined seed attributes. Heritability for the female-to-male flower ratio was over 100%, followed by yield (83.61) and plant height (87.73). The path analysis showed that number of branches (0.612), days from fruiting tomaturity (0.612), and the ratio of female to male flowers (0.789) had the strongest positive direct correlation with seed output (0.431). The number of days from flowering to fruiting had a negative indirect effect on yield. Ward's minimal variance cluster analysis' hierarchical clustering revealed phylo-geographic patterns in the genetic diversity. K-means clusteringrevealed that trees from different geographic regions weregrouped together in a cluster and as were trees from thesame geographical area placed in different clusters-suggestingthat geographical diversity did not go hand in handwith genetic diversity. In addition, clustering identifiedpromising accession with favourable traits for futureestablishment of elite seedling seed orchard and clonal seed orchard for varietal and hybridization programmes (**Table 1**).

Parameters	Minimum	Maximum	Mean	SD
Plant Height (cm)	24.30	257.30	152.69	80.81
C.D (cm)	6.0	24.5	15.65	6.04
No. of Branches	24.0	243.0	149.33	76.72
Crown width (cm)	124.00	383.80	276.66	93.51
Crown depth (cm)	142.33	262.33	215.73	42.58

Table 1.
Descriptive statistics of growth parameters of Jatropha curcas.

4.2 Morpho- and genetic-variability in *P. pinnata*

In order to evaluate genetic association and variability in seed and growth characters, 50 high yielding candidate plus trees (CPTs) of *P. pinnata* (L.) Pierre were collected from various locations within a latitudinal and longitudinal spread between 12°41 and 22°E longitude and 77°00 and 84°40N latitude, covering 11 locations in a spread of 150,000 km^2. Significant variations in plant height, number of branches, and seed form and oil content were also seen in the progeny trial. In comparison to seed traits, plant height and branch count showed substantially larger phenotypic and genotypic variance values. Oil content and seed length were the two seed parameters with the highest broad sense heritability (greater than 93% and 90.0%, respectively). In contrast, oil content displayed the largest genetic advance of 10.15%, while seed width displayed the second-highest genetic advance of 5.64%. By using Ward's Minimum Variance Cluster Analysis, phylo-geographic patterns of genetic diversity were revealed through hierarchical clustering. K means clustering showed that trees from the same geographic location were placed in distinct clusters, indicating that geographic diversity and genetic diversity are not correlated. Clustering also revealed prospective accessions with advantageous characteristics for the future establishment of orchards.

4.3 Evaluation trials of *J. curcas*

Ten elite lines of *J. curcas* (L.) vis-à-vis CRIDA-JJ-06, CRIDA-JL-06, CRIDA-JR-06, CSMCRI-C1, CSMCRI-C2, CSMCRI-C4, FRI-EL-1, NBPGR-0306, NBRI-J-05, and NBRI-J-18 were evaluated sequentially for 3 years (2007–2009) at CRIDA, Hyderabad, with the main objectives are to select superior plants with high seed and oil yields production and to study variation in agro-morphological, seed and oil yields characteristics. All attributes demonstrated significant variation among elite lines, according to analysis of variance. Although environment played a significant influence in the expression of these qualities, the broad sense heritability was high for all of the traits across years, showing that these traits were mostly regulated by genetic factors as opposed to environmental component. There were positive and highly significant correlations between the yield of seeds per plant and the yield of oil, pod weight, and pods per plant. Agro-morphological data-based cluster analysis was used to separate Jatropha lines into three groups using average linkage clustering. Clusters I, II, and III, respectively, included five, one, and four lines. The Jatropha populations showed highly significant genotypic differences for the variables investigated, and a strong genotype x environment interaction was observed for all traits. Jatropha lines differ

Genotypes	Seed yield (g/plant)	Oil yield (g/plant)	Pod Weight (g/plant)	East West Canopy	No. of primary branches	North South Canopy	Plant height (cm)	Pods / plant
CRIDAJJ06	449.36	143.82	520.86	195.90	79.35	195.52	200.50	225.79
CRIDAJL06	424.48	132.92	664.16	223.65	70.63	216.98	200.31	314.53
CRIDAJR06	434.63	138.20	540.92	205.15	82.06	221.25	202.87	285.25
CSMCRIC1	305.85	100.04	491.60	139.28	36.46	164.29	145.46	238.18
CSMCRIC2	327.06	101.97	443.82	201.54	69.50	216.52	177.31	198.09
CSMCRIC4	416.29	122.66	542.25	157.24	49.63	176.29	185.24	273.41
FRIEL1	435.30	153.90	550.81	193.54	73.25	221.95	196.86	323.09
NBPGR0306	410.67	133.38	575.79	163.42	90.50	172.47	199.54	230.80
NBRIJ05	320.99	83.13	478.28	183.13	81.48	194.17	213.29	206.33
NBRIJ18	288.04	111.45	413.24	163.19	94.48	159.69	186.14	207.87
Grand mean	381.27	122.15	522.17	182.60	72.73	193.91	190.75	250.33
CD (5%)	38.30	12.36	48.13	20.60	7.05	17.23	13.18	29.90
CV (%)	12.36	12.27	11.34	13.89	11.94	10.94	8.50	14.70

Table 2.
Comparative analysis of growth performance of elite lines of Jatropha curcas.

greatly from one another, so breeding attempts could employ these lines to improve the plant (**Table 2**).

Studies from other research groups on evaluation trials of *J. curcas* reported that both the phenotypic and genotypic coefficients of variation displayed a similar trend. 100-seed weight had the highest heritability (99%), followed by oil content (97%) and seed length (81%). Oil content and 100-seed weight showed a significant and positive connection (r = 0.517). Based on non-hierarchical K-Means cluster analysis, the *J. curcas* accessions were clearly divided into 6 clusters [18].

Prakash et al. [19] evaluated fifteen enzyme systems for their efficacy in distinguishing different accessions of *J. curcas* and it was found that twelve of them had fixed monomorphic alleles with no variation, while three (formate dehydrogenase, malate dehydrogenase, and peroxidase) were shown to be beneficial. The average number of polymorphic loci was 4, with a mean observed number of alleles per locus (A) of 1.533. Average observed heterozygosity (Ho) was 0.1082 and average anticipated heterozygosity (He) was 0.0993; gene flow (Nm) was 0.2177, indicating that there was little genetic variation between the different accessions, which may indicate that genes were not well segregated over generations.

Arolu [20] conducted similar evaluation trials in using 48 accessions of *J. curcas* from different locations of Malaysia and found that, with the exception of seed width, all of the morphological variables showed substantial levels of diversity and variation among the accessions. A positive and significant association between collar diameter, secondary branch count, and primary branch count was found. Additionally, a positive correlation between seed yield per ha, oil production per ha, and total number of seeds was found. With the exception of seed diameter and the number of major branches, high broad sense heritability was seen for all characteristics. The characteristic with the highest heritability was collar diameter (89.40%), and the one with the lowest value was seed width (−0.02%).

DOI: http://dx.doi.org/10.5772/intechopen.108793

4.4 Carbon sequestration potential and allometric equations

Periodic destructive sampling is one of the most accurate methods for measuring and tracking the above-ground biomass for a stand [12, 13]. Unfortunately, it takes a lot of time and effort to cut and weigh enough trees to accurately represent the size and species distribution in a system. Destructive harvesting is labour-intensive and cumbersome [16, 17]. Additionally, it is challenging to harvest trees destructively for research projects because they last longer and require ongoing monitoring of the trees being studied. In order to ascertain the above-ground biomass of different tree species used in agroforestry and forests, non-destructive approaches have been devised. These techniques are based on regression models that connect biomass to allometric growth factors [21, 22]. As they may be immediately linked to remotely sensed data for the calculation of biomass in wider regions, creating allometric connections using crown area and/or tree height as predictors of biomass is also gaining interest.

For the development of prediction equations, allometric equations in tree biomass are often coupled to easily observed predictor variables, such as collar diameter/DBH and tree height. An effective allometric equation predicts the biomass of a tree without the requirement of destructive sampling. A frequently used predictor is the diameter at ground level (collar diameter) or breast height [23–25]. According to Ghezehei et al. [25], a power function is typically employed to establish the relationship between several characteristics relevant to tree growth [26].

$$\text{Biomass} = \text{aX}^{\text{b}} \tag{1}$$

Where 'X'is a predictor (collar diameter, tree height etc.)

b is a scaling exponent or allometric coefficient and a is an intercept.

In this study, the correlations between collar diameter and biomass, tree height and biomass, and branch number and biomass are established using the aforementioned power function. The biomass, crown depth, and crown breadth were linearized in the aforementioned model. For total above, total below, and total (above + below) dry biomass, allometric equations were created. This methodwas employed to enable comparison with already-existing equations and to assess which equation yields more trustworthy findings. Using both sides of the logarithmic transformation.

$$\log\ (\text{biomass}) = (\log\ \text{a}) + \text{b}\ (\log\ \text{X}). \tag{2}$$

By applying linear regression in SPSS (version IBM SPSS Statistics 19), the parameters of the linearized allometric equations (b and log a) were computed. In the end, the antilog function was used to compute the original parameter "a". The R2 value, the F-statistic, and a scatter plot of the residuals were used to assess the significance and validity of the established equations. Microsoft Excel was employed to visualise the equation graphs. An independent dataset containing the data from 320 trees of varying ages and under diverse management settings was used to validate the proposed equations.

4.4.1 Carbon sequestration potential of J. curcas

One of our studies involved developing allometric relationships in the Jatropha plant to forecast several biomass-related components (above ground and below ground) using easily quantifiable characteristics, such as collar diameter, tree height, number of branches, crown diameter, and crown depth. Additionally, it was intended to demonstrate the validity of these associations using a separate dataset collected from a variety of managerial scenarios. In 2011, during the wet season, destructive sampling was done on Jatropha plants that were 8 years old. When predicting different biomass components (above, below, and total) using easily observable variables, highly significant allometric associations (F values significant at 1% level) were found with R^2values ranging from 0.89 to 0.98. Of all the predictors, collar diameter exhibited a highly significant relationship with total dry biomass per plant (R^2 = 0.97). The allometric relationships developed were validated with an independent dataset.

An independent dataset comprising of 220 trees with a suitable dendro-chronological range (2 years to 8 years) and management settings (various fertiliser and irrigation regimes) was used to validate the allometric connections. For the 220 trees, allometric relationships using collar diameter as the explanatory variable were utilised to forecast the number of branches. The number of branches-collar diameter allometric equation was found to be valid since the observed and predicted values show a high degree of agreement. Power regressions accurately depict tree allometry in a number of trees [27]. They enable the use of conventional least squares regression analysis and give uniformity of variance over the sampled range after log translation [25]. After the log transformation, a correction factor was applied in some research to lessen the systematic bias, and equations based on the correction factor were developed. However, because the computed F ratios are very significant and the R^2 values are greater than 0.90 at the 1% level of significance in our investigation, we did not utilise any correction factor. It implies that our study had little systematic bias. Estimates for some of the trees have become more accurate, thanks to the use of tree height and breast height diameter. However, as shown by collar diameter-based relationships, in the current study collar diameter alone was sufficient to reliably predict various tree development characteristics, including above ground, below ground, and total biomass.

In the current study, there is a consistent and significant allometric association between crown depth and the other growth characteristics. This went against several of the earlier findings [28]. Similar to this, animal browsing and seasonal deciduousness have an impact on some trees' biomass of leaves. In Hyderabad's semi-arid climate, we noticed that Jatropha trees lose their leaves in December and stay leafless from December through February. Beginning in the month of March, the leaves start to grow. Therefore, during periods when leaf drop does not occur, the equations for the estimate of leaf biomass are to be applied (**Table 3**).

Diameter Classes	Above dry ground dry biomass t/ha	Below ground dry biomass t/ha	Total dry t/ha (A + B)	Total C biomass t/ha (a + b)
D1 (5–10 cm)	3.10	2.23	5.33	2.56
D2 (10–15 cm)	14.28	4.28	18.56	7.84
D3 (15–20 cm)	16.19	5.32	21.51	9.12
D4 (20–25 cm)	18.44	5.56	24.00	10.08
Average	13.00	4.35	17.35	7.36

Table 3.
Jatropha curcas *dry biomass and carbon biomass.*

4.4.2 Evaluation trials of P. pinnata

In a study, Sharma et al. [29] investigated the molecular genetic diversity of 46 accessions of *P. pinnata* gathered from six different Indian states. They discovered that 520 identifiable fragments, of which 502 (96.5%) were polymorphic, were produced by 5 AFLP primer combinations. Four factors—polymorphism information content (PIC), effective multiplex ratio (EMR), marker index (MI), and resolving power—were evaluated to determine how informative an AFLP primer is (RP). A total of 51 distinct fragments were found, of which 19 distinct fragments were found using the primer pair E-ACG/M-CTA. Even though the neighbour joining (NJ) method did not strictly classify accessions according to the region in which they were collected, a high level of genetic diversity was found in the tested germplasm.

Patil et al. [30] conducted a comparative study on the biodiesel and bioproductive parameters of *J. curcas* and *P. pinnata*. The variation of bioproductivity and biodiesel parameters of both plants were compared every 6 months for years and analysed with variance and correlation coefficient by Pearson's method using Graphpad instat 3.06 software (Windows and Mac). *P. pinnata* has better germination rate (71%), 100 pod weight (PW) (311.59 g) and 100 seed weight (SW) (173.6 g) compared to *J. curcas* germination rate (3.2%), 100 pod weight (111.29). g) and 100 SW (67.6 g). *P. pinnata* has a strong correlation between plant height and crown growth (CG) (0.98), neck diameter (CD) (0.99), number of branches per plant (NBP) (0.995) and number of leaves per branch (NLB) (0.862) compared to *J. curcas* which showed a good correlation between plant height and CG (0.976), CD (0.970), NBP (0.988), NLB (0.920) and number of pods per branch (0.657). However, *J. curcas* presented a negative correlation between beam width and seed length (SL) (−0.7), seed width (−0.28) and seed thickness (ST) (−0.36) and 100 PW and SL (−0.199). ST (−0.220) and 100 SW (−0.70). For each litre of crude oil, approximately kg of *P. pinnata* seeds were consumed, yielding 896 mL of biodiesel when trans-esterified, compared to 5.66 kg of *J. curcas* seeds per litre of crude oil, yielding approximately 663 mL of biodiesel.

4.4.3 Carbon sequestration potential of P. pinnata

Additionally, the study established allometric relationships in *P. pinnata* between biomass and the readily measurable characteristics (collar diameter, tree height, number of branches, crown diameter, and depth), and it tested the validity of these relationships using a separate dataset from various management scenarios. Plants were 9 years old when destructive sampling was done in the months of January through March 2012.

Using the collar diameter ($R^2 > 0.96$), plant height ($R^2 > 0.94$), number of branches ($R^2 > 0.91$), crown width ($R^2 > 0.96$), and crown depth ($R^2 > 0.65$), highly significant allometric associations (F values significant at 1% level) were developed. An independent dataset was used to validate the associations that were developed. Independent datasets from 320 trees with diverse age group (2 years to 9 years) and management (different fertilisation, irrigation, grafting, and seedling generated plants) circumstances were used to assess the allometric relationships. The correlation coefficient was high and positive ($R^2 > 0.917$ and $R^2 > 0.978$), indicating that the link was linear. As the independent data sets comprised a variety of ages, management situations, and development settings, the results substantially support the validity of the height-collar diameter and number of branches-collar diameter equations (**Table 4**).

Diameter Classes	Above dry ground dry biomass t/ha	Below ground dry biomass t/ha	Total dry t/ha (A + B)	Total C biomass t/ha (a + b)
D1 (0–5 cm)	2.17	2.16	4.33	1.87
D2 (5–10 cm)	11.36	11.40	22.76	9.80
D3 (10–15 cm)	27.79	22.45	50.24	21.72
D4 (15–20 cm)	42.31	38.49	80.81	34.85
Average	20.91	18.63	39.54	17.06

Table 4.
P. pinnata *dry biomass and carbon biomass.*

For a number of tree species, species-specific allometric equations have been devised [31, 32]. Such Jatropha equations, however, are limited and unavailable, especially for Indian circumstances. Allometry places a lot of focus on quantifying above-ground biomass rather than below-ground biomass because the latter is laborious and time-consuming. There is a steady requirement to create allometric equations of Jatropha because it is suggested that the species be widely planted in semi-arid regions of India. In the absence of site-specific or generalised equations, equations created elsewhere may be taken into consideration for usage at a site [33]. However, for Jatropha, there are very few published equations, and even fewer for the quantification of below-ground biomass. For trees like Jatropha, the development of allometric correlations is particularly difficult due to their complicated form-function relationships, defoliation, and change in leaf phenology during dry seasons, which is especially noticeable in younger trees [34]. There is a knowledge gap regarding accurate and thorough estimates of Jatropha biomass, despite new scientific knowledge on utility features [35], genetic diversity assessment [36], and yield [37] being developed. As a result, research was done with the aim of developing allometric equations and validating them using a large, independent dataset that represented various management settings.

5. Successful agroforestry models

5.1 Pongamia: *Macrotyloma uniflorum* (horse gram) based agroforestry model

A study utilising *P. pinnata* as main crop and *M. uniflorum* (Horse gram) as intercrop were carried out to know the yield patterns in these two crops in varied spacings. *P. pinnata* was sown in 2003 in 3 different spacings of 6 × 4 m, 6 × 6 m and 6 × 8 m, though 6 × 6 m is general recommendation, which was used as the check for the study. In 2007 (at the end of 4th year of plantation), Kernel yield of *P. pinnata* in recommended spacing (check) of 6 × 6 m was found to be high (0.38 kg/tree) though the intercropped *M. uniflorum* yield is minimum (2.9 q/ha) as compared to spacing of *P. pinnata* with 6 × 4 m which yielded 0.01 kg/tree that translates to97.4% and in 6 × 8 m which yielded 0.02 kg/tree or about 94.7% as compared to recommended spacing (check). The intercropped yield of Horse gram in 2007 was 3.3 and 4.3 q/ha under 6 × 4 m and 6 × 8 m spacing respectively showing an increase of 13.8% and 48.3% over the yields of *M. uniflorum* obtained in 6 × 6 m (2.9 q/ha). However, in 2008 also like in 2007, the kernel yield of *P. pinnata* in recommended spacing was the highest (1.88 kg/tree) as compared to 6 × 4 m which yielded 1.02 kg/tree ie. −45.7%

while in 6 × 8 m it yielded 0.18 kg/tree ie. −90.4% as compared to recommended spacing (check). These results indicate the impact of intercropping under recommended spacing for enhanced yield of *P. pinnata*. The area (ha), production (ha) and productivity (kg/ha) of *P. pinnata* in India are 1702.9, 719.5 and 423 respectively while in the state of Andhra Pradesh in India, these are 272.5, 76.3 and 280 respectively. *M. uniflorum* is also a bankable contingent crop. It is beneficial in curing cough, breathing illness arising due to flatulation, hiccups, stones and fever. It also helps eliminate germs and worms from the body.

Both the main crop *P. pinnata* as well as the intercrop *M. uniflorum* are drought tolerant legumes with nitrogen fixing ability to the soil. Hence the study underscored that an agroforestry model with a tree legume and a crop legume would be very beneficial in dry lands and wastelands as land reclamation and rejuvenation happens through soil-enriching legumes. However, to balance between the food cropped area with these biofuels the best option would be intercropping for sustaining the nutrition security of human and animal populations along with the biofuel oils within a given area. This intercropping wouldalso serve as a risk aversion strategyin providing nutritional security and/or as a crop of economic value during the early stages of the growth of *P. pinnata*. Since the full potential growth of *P. pinnata* takes 7 years, intercropping especially with pulses would benefit the main crop through means ofroot nodulation, and leaf fall thereby enhancing the soil fertility.

The above results on intercropping with legume based biofuel tree and grain legume may be recommended in arid and semi-arid drought prone areas, watersheds and wastelands for sustainable realisation of the renewable source of energy through the biofuel oil yielding tree *P. pinnata* along with the arid and semiarid grain legume *M. uniflorum* for sustaining the nutritional security of animals and humans.

5.2 *J. curcas*: *Cajanus cajan* (Redgram) based agroforestry model

Three best genotypes (Jabua, CRIDA-Utnoor& Raipur) based on seed yield and oil content in seed were selected from a total of 102 genotypes were selected for the study. These accessions were collected from the states of Andhra Pradesh, Madhya Pradesh and Chhattisgarh during 2002–2003 and planted in 2003 July at Research farm of CRIDA, Hyderabad.

The date of sowing of main crop *J. curcas* was 18.09.2003. With the spacing of 3 m × 3 m a population of 1111 plants was accommodated in each hectare. Total plot size under Jabua, CRIDA-Utnoor & Raipur varieties of Jatropha with 3 m × 3 m was 1620 m^2, 2160 m^2 and 2160 m^2 respectively. Thus the total plot size under main crop was 5940 m^2. Number of plants accommodated under Jabua, CRIDA-Utnoor and Raipur were 180, 240 and 240 respectively. Thus the total plant population under these three varieties was 660. *C. cajan* was taken as intercrop in all the three genotypes during 2004. Sowing of *C. cajan*, variety ICPL-85063 (Lakshmi) was done on 14th July during 2004 with spacing of 60 cm × 10 cm. Number of plants of *J. curcas* and *C. cajan* per hectare area was 1111 and 1667 respectively. At the time of sowing fertilisers of 40 kg N and 50 kg $P_2 0_5$ per hectare were applied. Harvesting of *C. cajan* was done on 20th November 2004.

The primary treatments of pruning were effected in January 2004, which was dormant stage. In each of the three test genotypes of *J. curcas*, three ground level pruning treatments viz. 30 cm, 45 cm and 60 cm were affected along with the unpruned control. The seed yields of the control as well as the three pruning treatments were recorded during 2005, 2006 and 2007.

The study reaffirmed the fact that Jatropha is an easy to establish crop, grows relatively quickly and is hardy and drought tolerant [38]. Highest yield of legume intercrop was obtained in CRIDA-Utnoor genotype of *J. curcas*, which was also the highest seed yielder. Thus *J. curcas–C. cajan* based Agroforestry Model may be recommended in semi-arid to arid regions that can be brought under eco restoration efforts not only for biodiesel production but also to increase global green cover.

6. Sustainability model for addressing rural energy needs

6.1 Ecosystem subsidies

Nature has its own ways that determine the flow of nutrients, energy and organic matter among different ecosystems. All such resource flows which augment the population of consumers in the ecosystem are called ecological subsidies. Conventional sources such as fossil fuels and curated fuels such as petrol and diesel depreciate the natural energy flow of any ecosystem by polluting the environment and clogging sensitive physiological receptors of plants while biofuels present a sustainable alternative. *J. curcas* and *P. pinnata* plantations add large amount of organic matter through fallen leaves and biomass, it helps to improve the microbial biological activity hence catalysing the flow of nutrients between different ecosystems.

6.2 Restoration of wastelands

Restoration of degraded lands and combating desertification have been in the forefront of national priorities since 1980s, the significance of which have mounted due to inclement weather, climate crisis and the imminent threat of global warming. In 2003–2005, 94.53 million hectares (mha) of land underwent land degradation, which gradually increased to 96.40 mha in 2011–2013 and 97.85 million hectares in 2018–2019. Main reasons attributed to the degradation include loss of soil cover, vegetation loss, and wind and water erosion taken atop by climate crisis. India aims to restore 26 million hectares of degraded land by 2030 and is also working towards achieving its national commitment on Land Degradation Neutrality (LDN). National Afforestation Programme implemented since 2000, National Action Programme to Combat Desertification (2001), National Mission on Green India and commitment laid towards Bonn's Challenge calls for immediate intervention to elbow out the crisis of land degradation.

Results of a study on carbon sequestration potential and greenhouse gas emissions in *J. curcas* and *P. pinnata* show that these TBOs grown on dry lands have the potential to add large quantities of carbon (0.3% - 1.7% in *J. curcas* and 0.3% - 0.48% in *P. pinnata*) in addition to that accumulated by the standing crop. Increased carbon content in the soil augment the growth of beneficial microbes which in turn improves the soil fertility, productivity and ecological subsidies of the landscape.

6.3 Reducing carbon emission and working towards climate goals

Recently, at the 26th Conference of Parties (COP 26) Climate Summit held at Glasgow, India pledged to attain carbon neutrality by 2070 as part of a 5-point action plan that included reducing emissions to 50 percent by 2030. The role of vegetation (terrestrial and aquatic) in sequestering carbon and thereby reducing the emissions are widely acknowledged and taken forward through afforestation programmes. Certain plants such as *J. curcas* and *P. pinnata* have a higher rate of sequestration of carbon. Research shows that a nine-year-old *J. curcas* and *P. pinnata* can respectively sequester 7.36 tonnes/hectare and 17.06 tonnes/hectare of carbon per annum (my paper reference). Carbon di oxide mitigation potential of a nine-year-old *J. curcas* and *P. pinnata* respectively accounted to 31.54 tonnes/hectare and 50.18 tonnes/hectare, which exemplifies the role of *J. curcas* and *P. pinnata* in mitigating the effects of a major greenhouse gas. Adoption of *P. pinnata* and *J. curcas* either as monoculture plantations or as components in agroforestry models not only support biodiesel production but aid in our attempt to cut down emissions and achieve climate goals within the pledged deadline.

7. Conclusion

The policies for encouraging the plantation of tree-based biofuels are focused on the use of wastelands and less productive regions. In light of stagnant food crops production and burgeoning domestic demand for food grains, use of food crops for biofuels production remains as question, as it competes with the crops for scare land and water resources. Additionally, India's policy of neither allowing nor encouraging the use of food or other feedstock for the production of biofuels is based on the fact that food inflation has been steadily rising over the past few years. As a result, the nation's biofuel project must carefully navigate through sensitive local and international issues. Climate Financing is a major hurdle caught between bureaucratic red tapism and false promises made by developed countries. Without adequate finance and policy support, technologies cease to transfer from lab to land. India also urged the developed countries to deliver on their promises of climate financing at the recently concluded COP 26 Climate Summit at Glasgow. Biofuel subsidies are already in place in United States and European Union to facilitate farmers facing low prices for their crops. Subsidies in the form of incentives need to be provided for start-ups that venture into biofuel crops farming, biofuel production and supply. Such incentives not only boost up the green start-ups but also motivates entrepreneurs to invest in such businesses which synergizes our effort to attain carbon neutrality by 2070.

Author details

Rajeshwar Rao Gandhe*, Ajin Sekhar and Rajkumar Muthu
Tropical Forest Research Institute, Jabalpur, Madhya Pradesh, India

*Address all correspondence to: grrcrida@gmail.com

References

[1] Heller J. Physic nut, *Jatropha curcas* L. Promoting the Conservation and Use of Underutilized and Neglected Crops. 1. Gatersleben, Rome: Institute of Plant Genetics and Crop Plant Research, International Plant Genetic Resources Institute; 1996

[2] Openshaw K. A review of Jatropha curcas: An oil plant of unfulfilled promise. Biomass and Bioenergy. 2000; **19**(1):1-5

[3] Zahawi. Establishment and growth of living fence species: An overlooked tool for the restoration of degraded areas in the tropics. Restoration Ecology. 2005; **13**(1):92-102

[4] Gubitz GM, Mittelbach M, Trabi M. Exploitation of the tropical oil seed plant *Jatropha curcas* L. Bioresource Technology. 1999;**67**(1):73-82

[5] Hamilton K, Bayonr R, Turner G, Higgins D. State of the Voluntary Carbon Markets: Picking Up Steam. Washington, DC: New Carbon Finance, London and The Ecosystem Marketplace; 2007

[6] Spaan W, Bodnar F, Idoe O, Graaff JD. Implementation of contour vegetation barriers under farmer conditions in Burkina Faso and Mali. Quarterly Journal of International Agriculture. 2004;**43**(1):21-38

[7] Cano-Asseleih LM, Plumbley RA, Hylands PJ. Purification and partial characterization of a hemagglutinin from seeds of Jatropha curcas. Journal of Food Biochemistry. 1989;**13**(1):1-20

[8] Foidl N, Kashyap A. Exploring the Potential of Jatropha curcas in Rural Development and Environmental Protection. New York: Rockefeller Foundation; 1999. p. 184

[9] Jongschaap REE, Corré WJ, Bindraban PS, Brandenburg WA. Claims and facts on *Jatropha curcas* L.: Global *Jatropha curcas* evaluation. In: Breeding and Propagation Programme (No. 158). Wageningen, Netherlands: Plant Research International; 2007

[10] Bringi NV. Non-traditional Oilseeds and Oils in India. New Delhi: Oxford and IBH Publishing Co; 1987

[11] Sreedevi TK, Wani SP, Osman M, Singh SN. Participatory research and development to evaluate Pongamia seed cake as source of plant nutrient in integrated watershed management. Journal of SAT Agricultural Research. 2009;7(1):13-14

[12] Norries DN, Blair JM, Johnson LC, McKane RB. Assessing changes in biomass, productivity, and carbon stores following *Juniperus virgiana* forest expansion into tall grass prairie. Canadian Journal of Forest Research. 2001;**31**:1940-1946

[13] Van TK, Rayachhetry MB, Centre D. Estimating aboveground biomass of *melaleuca quinquenenervia* in Florida. USA Journal Aqua Plant Management. 2000;**38**:62-67

[14] Wadham-Gagnon B, Sharpe D. Estimating Carbon Stocks in Tropical Hardwood Plantations: Using Species-Specific and Non-destructive Parameters to Estimate Aboveground Biomass for Six Native Species in Panama. Internship Report. Panama, Republic of Panama: Smithsonian Tropical Research Institute ENVR; 2006. p. 451

[15] Delitti WBC, Meguro M, Pausas JG. Biomass and mineral mass estimates in a

"cerrado" ecosystem. Revista Brasil Bot. 2006;**29**(4):531-540

[16] Kale M, Sing S, Roy PS, Desothali V, Gholem VS. Biomass equations of dominant species of dry deciduous forests in Shivupuri district, Madhya Pradesh. Current Science. 2004;**87**(5): 683-687

[17] Telenius B, Verwijs T. The influence of allometric variation, vertical biomass distribution and sampling procedure on biomass estimates in commercial short rotation forests. Bioresource Technology. 1995;**51**:247-253

[18] Kumar R, Das N. Survey and selection of *Jatropha curcas* L. germplasm: Assessment of genetic variability and divergence studies on the seed traits and oil content. Industrial Crops and Products. 2018;**118**:125-130

[19] Prakash MS, Warrier RR, Anandalakshmi R. Assessment of genetic diversity of *Jatropha curcas* (l.) in India. Coimbatore, India: Institute of Forest Genetics and Tree Breeding; 2008

[20] Arolu IW. Genetic Diversity and Heritability Estimate of *Jatropha Curcas* L. Accessions Using Agro-Morphological and Molecular Markers [Thesis]. Selangor Darul Ehsan, Malaysia: Universiti Putra Malaysia; 2012

[21] FAO. Estimating Biomass and Biomass Change of Tropical Forests: A Primer. Rome: FAO; 1997

[22] Lott JE, Howard SB, Black CR, Ong CK. Allometric estimation of above-ground biomass and leaf area in managed *Grevillea robusta* agroforestry systems. Agroforestry Systems. 2000; **49**:1-15

[23] Achten WMJ, Maes WH, Reubens B, Mathijs E, Singh VP, Verchot L, et al. Biomass production and allocation in *Jatropha curcas* L. seedlings under different levels of drought stress. Biomass and Bioenergy. 2010; **34**(5):667-676

[24] Firdaus MS, Hanif AHM, Safiee AS, Ismail MR. Carbon sequestration potential in soil and biomass of *Jatropha curcas*. In: Presented at: 19th World Congress of Soil Science, Soil Solutions for a Changing World. Brisbane, Australia: Published on DVD; 2010. pp. 1-6

[25] Ghezehei SB, Annandale JG, Everson CS. Shoot allometry of *Jatropha curcas*. Southern Forest. 2009;**71**(4):279-286

[26] Ong JE, Gong WK, Wong CH. Allometry and partitioning of the mangrove, Rhizophoraapiculata. Forest Ecology and Management. 2004;**188** (1–3):395-408

[27] Clough BF, Scott K. Allometric relationships for estimating above-ground biomass in six mangrove species. Forest Ecology and Management. 1989; **27**:117-127

[28] Ter-Mikaelian MT, Korzukhin M. Biomass equations of sixtyfive north American tree species. Forest Ecology and Management. 1997;**97**:1-24

[29] Sharma SS, Negi MS, Sinha P, Kumar K, Tripathi SB. Assessment of genetic diversity of biodiesel species Pongamiapinnata accessions using AFLP and three endonuclease-AFLP. Plant Molecular Biology Reporter. 2011;**29**(1): 12-18

[30] Patil VK, Bhandare P, Kulkarni PB, Naik GR. Progeny evaluation of *Jatropha curcas* and Pongamiapinnata with comparison to bioproductivity and biodiesel parameters. Journal of Forestry Research. 2015;**26**(1):137-142

[31] Levia DF. A generalized allometric equation to predict foliar dry weight on the basis of trunk diameter for eastern white pine (*Pinusstrobes* L.). Forest Ecology and Management. 2008;**255**: 1789-1792

[32] Wang JR, Letchford T, Comeau P, Kimmins JP. Above and below-ground biomass and nutrient distribution of a paper birch and subalpine fir mixed species stand in the sub-boreal spruce zone of British Columbia. Forest Ecology and Management. 2000;**130**:17-26

[33] Pastor J, Abet JD, Melillo J. Biomass prediction using generalized allometric regressions for some northeast tree species. Forest Ecology and Management. 1984;**7**:265-274

[34] Cole TG, Ewel JJ. Allometric equations for four valuable tropical tree species. Forest Ecology and Management. 2006;**229**(1–3):351-360

[35] Gunaseelan VN. Biomass estimates, characteristics, biochemical methane potential, kinetics and energy flow from *Jatropha curcas* on dry lands. Biomass and Bioenergy. 2009;**33**:589-596

[36] Rao GR, Prasad YG, Prabhakar M, Srinivas I, Rao KV, Korwar GR, et al. Biofuel Crops for Drylands: Cultivation and Processing Issues, Technical Bulletin 1/2010. Hyderabad, India: Central Research Institute for Dryland Agriculture Hyderabad; 2008. p. 32

[37] Lapola DM, Priess JA, Bondeau A. Modelling the land requirements and potential productivity of sugarcane and *Jatropha* in Brazil and India using the LPJmL dynamic global vegetation model. Biomass and Bioenergy. 2009;**33**: 1087-1095

[38] Behera SK, Srivastava P, Tripathi R, Singh JP, Singh N. Evaluation of plant performance of *Jatropha curcas* L. under different agro-practices for optimizing biomass–a case study. Biomass and Bioenergy. 2010;**34**(1):30-41